MAPPING DOGGERLAND
The Mesolithic Landscapes of the Southern North Sea

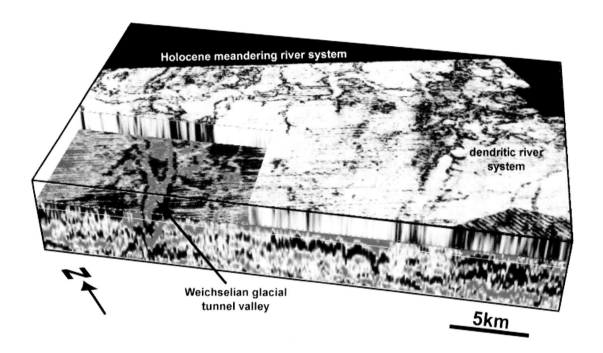

Edited by

Vincent Gaffney, Kenneth Thomson and Simon Fitch

A project funded by the Aggregates Levy Sustainability Fund
and administered by English Heritage

Archaeopress

Published by

Archaeopress
Publishers of British Archaeological Reports
Gordon House
276 Banbury Road
Oxford OX2 7ED
England
bar@archaeopress.com
www.archaeopress.com

MAPPING DOGGERLAND
The Mesolithic Landscapes of the Southern North Sea

© **Institute of Archaeology and Antiquity, University of Birmingham 2007**

ISBN 978 1 905739 14 1

All rights reserved. No part of this publication may be reproduced, stored in retrieval system, or transmitted, in any form or by any means, electronic, mechanical, photocopying or otherwise, without the prior written permission of the copyright owner.

Printed in England by Alden HenDi, Oxfordshire

In Memoriam

Dr Kenneth Thomson
(1966-2007)
Pioneer and Explorer of the North Sea

Contents

Foreword	Huw Edwards (PGS)
Preface	Ian Oxley (English Heritage)
Acknowledgements	

1. Mapping Doggerland

Vincent Gaffney and Kenneth Thomson

1.1	Introduction	1
1.2	The context of study	1
1.3	Previous methodological approaches	4
1.4	Towards an alternative methodology	6

2. Coordinating Marine Survey Data Sources

Mark Bunch, Vincent Gaffney and Kenneth Thomson

2.1	Introduction	11
2.2	Identification of sources: who acquires, owns or holds survey data?	11
2.3	Key data repositories	11
2.4	Governance and surveying within territorial waters	13
2.5	Public metadata resources	14
2.6	Case study assessments of survey data	16
2.6.1	Area 1: the Bristol Channel	16
2.6.2	Area 2: Portland	21
2.6.3	Area 3: The Spurn	21
2.7	Conclusions	21

3. 3D Seismic Reflection Data, Associated Technologies and the Development of the Project Methodology

Kenneth Thomson and Vincent Gaffney

3.1	Introduction	23
3.2	Seismic reflection method and resolution	23
3.3	2D versus 3D seismic acquisition and interpretation	27
3.4	Interpretation strategy for the Southern North Sea	28
3.5	Conclusions	30

4. Merging Technologies: The integration and visualisation of spatial data sets used in the project

Simon Fitch, Vincent Gaffney and Kenneth Thomson

4.1	Introduction	33
4.2	Infrastructure	34
4.3	Software integration	35
4.4	Primary integration procedures	35
4.5	Integration of volumetric information through solid modelling	36
4.6	Merging technologies	40
4.7	Conclusions	41

5. A Geomorphological Investigation of Submerged Depositional Features within the Outer Silver Pit, Southern North Sea

Kate Briggs, Kenneth Thomson and Vincent Gaffney

5.1	Introduction	43
5.2	Feature descriptions	44
5.3	Discussion/Feature classification	50
	5.3.1 Temperate terrestrial geomorphological processes	51
	5.3.2 Marine geomorphological processes	51
	5.3.3 Subaerial coastal landforms	52
	5.3.4 Marine bedforms	53
5.4	Environmental Interpretation	55
	5.4.1 Sand Bank classification	55
	5.4.2 Estuaries and Sand Bank Formation	56
5.5	Conclusions	58

6. Salt Tectonics in the Southern North Sea: Controls on Late Pleistocene-Holocene Geomorphology
Simon Holford, Kenneth Thomson and Vincent Gaffney

6.1	Introduction	61
6.2	Relationships between salt structures and late Pleistocene-Holocene fluvial systems.	61
6.3	Conclusions	65

7. An Atlas of the Palaeolandscapes of the Southern North Sea
Simon Fitch, Vincent Gaffney, Kenneth Thomson
with Kate Briggs, Mark Bunch and Simon Holford

7.1	Introduction	67
7.2	North Western Quadrant	72
	7.2.1 Description	72
	7.2.2 Other Features	75
	7.2.2.1 Solid Geology	75
	7.2.2.2 Recent Geological Features	75
7.3	North Eastern Quadrant	75
	7.3.1 Description	75
	7.3.2 Other Features	81
	7.3.2.1 Solid Geology	81
	7.3.2.2 Recent Geological Features	81
7.4	South East Quadrant	81
	7.4.1 Description	81
	7.4.2 Other Features	82
	7.4.2.1 Recent Geological Features	82
7.5	South Western Quadrant	86
	7.5.1 Description	86
	7.5.2 Other Features	89
	7.5.2.1 Fluvio-Glacial features	89
	7.5.2.2 Recent Geological Features	89
7.6	Conclusions	89

8. The Potential of the Organic Archive for Environmental Reconstruction: An Assessment of Selected Borehole Sediments from the Southern North Sea.

David Smith, Simon Fitch, Ben Gearey, Tom Hill,
Simon Holford, Andy Howard and Christina Jolliffe

8.1	Introduction	93
8.2	Potential and Rationale	93
8.3	Core Selection	97
8.4	**Palaeoenvironmental Assessment**	97
	8.4.1 Sampling	97
	8.4.2 Visual Assessment	97
	8.4.3 Assessment of macrofossil (insect and plant) inclusions	97
	8.4.4 Pollen Assessment	97
	8.4.5 Results	99
	8.4.5.1 Vibrocore 53/02/395 (Figure 8.2)	99
	8.4.5.2 Vibrocore 54/02/215 (Figure 8.3)	101
	8.4.5.3 Vibrocore 54/02/80 (Figure 8.4)	101
	8.4.5.4 Borehole 81/50	101
8.5	Discussion	101
8.6	Conclusions	101
Appendix		103

9. Heritage Management and the North Sea Palaeolandscapes Project

Simon Fitch, Vincent Gaffney and Kenneth Thomson

9.1	Introduction	105
9.2	Future Research	108
9.3	Cultural resource management procedures in the Southern North Sea	110
9.4	Landscape Characterisation	110
9.5	Threat mapping	116
9.6	Threat and Uncertainty Mapping	116
9.7	Final Observations	116
Bibliography		119
Abstract		127
Index		129

List of Figures

Figures

Figure 1.1 Hypothetical maximum extent of Doggerland 2
Figure 1.2 Early Holocene Doggerland 3
Figure 1.3 Holocene shorelines 5
Figure 1.4 Current extent of Southern North Sea Megasurvey 3D seismic data 7
Figure 2.1 BGS survey coverage on the UK continental shelf 12
Figure 2.2 Distribution of 2D and 3D seismic surveys acquired by the UKOOA 14
Figure 2.3 Three study areas around the English coast 15
Figure 2.4 Industrial threats to seabed archaeology within the Bristol Channel 17
Figure 2.5 Distribution of BGS and UKOOA surveys within Area 2 18
Figure 2.6 Area 3 19
Figure 2.7 Intense use of space within the Southern North Sea 20
Figure 3.1 Typical marine seismic reflection acquisition 24
Figure 3.2 Seismic resolution of a layer of varying thickness 24
Figure 3.3 Plots of seismic resolution as a function of burial depth and frequency 25
Figure 3.4 A comparison between high frequency 2D seismic reflection line and low frequency 3D seismic lines 26
Figure 3.5 Four possible interpretations of a channel morphology based on a coarse 2D seismic grid 27
Figure 3.6 Typical 3D marine seismic reflection acquisition 27
Figure 3.7 Poole and Christchurch bays 29
Figure 3.8 A comparison of seismic images from the Dogger Bank produced using different techniques 29
Figure 4.1 HP VISTA infrastructure diagram 33
Figure 4.2 Inspecting data in stereo at the Visual and Spatial Technology Centre 34
Figure 4.3 Seismic data slice showing fluvial channel and estuary 35
Figure 4.4 3D amplitude surface within a GIS 37
Figure 4.5 Segmentation of features of interest from a seismic volume 37
Figure 4.6 Wrapping of identified features within seismic data 38
Figure 4.7 Solid model generated by wrapping 39
Figure 4.8 Removal of elements of the solid model within the Avizo package 39
Figure 4.9 Exported solid model within the GIS system 40
Figure 4.10 GIS layers display within a fully 3 dimensional environment 41
Figure 5.1 40 and 60m bathymetric contours of the Southern North Sea 43
Figure 5.2 Bathymetric contours of the Outer Silver Pit area 44
Figure 5.3 Hilbert transform time slice at 0.06 seconds 45
Figure 5.4 3-D illuminated view of Ridges A and B 46
Figure 5.5 Dimensions of the OSP and Ridges A and B 47
Figure 5.6 Two 2D seismic lines running through Ridge A and their location 48
Figure 5.7 Quaternary Geology map of the eastern end of the OSP 49
Figure 5.8 RMS amplitude map of the eastern end of the OSP 50
Figure 5.9 Truncated strata on the bed of the Outer Silver Pit 52
Figure 5.10 Changes in land area with rising sea level based upon the depth to base Holocene map 54
Figure 5.11 Postulated land configuration at the time when the OSP sand banks were last active 57
Figure 6.1 Time slices centred on prominent salt dome 62
Figure 6.2 Collapse graben, salt swell and fluvial channel 63
Figure 6.3 BGS sparker profile 81/03/53 64
Figure 7.1 An RMS timeslice covering the whole of the project study area 68
Figure 7.2 Primary features identified within the Holocene landscape of the southern North Sea 69
Figure 7.3 Pre-Holocene features recorded during mapping 70
Figure 7.4 General map of all recorded Holocene landscape features including general topographic interpretation 71
Figure 7.5 The Holocene landscape and features within the northwestern quadrant 72
Figure 7.6 Vertical slice through salt dome exhibiting graben collapse 73
Figure 7.7 Major fluvial channel deviating around an underlying salt structure 73
Figure 7.8 Seismic relief image of the Flamborough head disturbance 74
Figure 7.9 Western end of the Outer Sliver Pit lake showing outflow channel 76
Figure 7.10 A seismic line across the Outer Silver Pit 77
Figure 7.11 The Holocene landscape and features within the northeastern quadrant 78
Figure 7.12 Junction of rivers and coastline 79

Figure 7.13 A series of Tunnel Valleys crossing the Outer Silver Pit 79
Figure 7.14 Location of a small structure that resemble a palaeochannel 80
Figure 7.15 Modern Sandwaves directly overlying the Holocene landscape 81
Figure 7.16 The Holocene landscape and features within the southeastern quadrant 83
Figure 7.17 Seismic line across the southeastern quadrant 84
Figure 7.18 Representative image of the "mottling" within the seismic data 84
Figure 7.19 Cross section over Markham's Hole (BGS line 80-01-05) 85
Figure 7.20 Seismic timeslice of the area interpreted as a salt marsh 85
Figure 7.21 The Holocene landscape and features within the southwestern quadrant 87
Figure 7.22 Cross section through Well Hole (BGS Line 93-01-81) 88
Figure 7.23 Cross section through the large channels in the southwestern quadrant. (BGS Line 93-01-74A) 89
Figure 7.24 Image showing the complex structure of a glacial outwash plain 89
Figure 8.1 Location of the boreholes in the area included in the 3D seismic survey 94
Figure 8.2 Location of the boreholes requested from the area and primary features identified during mapping 95
Figure 8.3 Location of the boreholes inspected from the area included with the 3D seismic survey 96
Figure 8.4 Pollen diagram for Vibrocore 53/02/395 99
Figure 8.5 Pollen diagram for Vibrocore 54/02/215 100
Figure 8.6 Pollen diagram for Vibrocore 34/02/80 100
Figure 9.1 Major topographic or economic zones within the study area 107
Figure 9.2 Probable late Palaeolithic land surfaces adjacent to the Norwegian trench 109
Figure 9.3 Seismic data cube illustrating chronostratigraphic relationship between Holocene and earlier features 109
Figure 9.4 The analytical process 111
Figure 9.5 Broad landscape character zones 113
Figure 9.6 Cross correlation of major topographic and landscape characterisation zones 114
Figure 9.7 Potential for preservation 115
Figure 9.8 Red flag mapping 117

Tables

Table 5.1 Position, dimensions and trends of Ridges A and B 44
Table 5.2 Average dip of the flanks of Ridges A and B 46
Table 7.1 Basic quantitative data relating to identified landscape features 90
Table 8.1 Summary of sedimentary samples and environmental assessment undertaken from the Southern North Sea vibrocores and boreholes 98
Table 9.1 Primary landscape characterisation zones 111
Table 9.2 Ranking of features by relative archaeological potential 116

Foreword

"The middle of the North Sea? Mammoth tusks and flint spears! Looking for Doggerland?   All we need is several million dollars worth of your 3D seismic data!"

I am often asked to become part of many research projects, but this phone call from Birmingham was outlining one of the most intriguing. The voice on the phone said "We will come to visit and explain what we want to try to do".

So a few weeks later, we huddled round one of our computer workstations and instead of looking deep down in the seismic data for oil, we applied the latest in petroleum exploration technology to the shallow section. To our amazement, for the first time in thousands of years, the long forgotten surface of Doggerland started to appear. Science does not get more exciting than this and it dawned on us that we were witnessing the start of a new era in marine archaeology.

The project brought together a wide range of people, interests, expertise and technology. The project's achievements, which are of international significance, are a credit to Birmingham University, the team and sponsoring companies for which we are proud to have received a British Archaeological Award. These achievements also stand as a lasting tribute to my friend and colleague Dr Ken Thomson, who tragically died following the project conclusion. His infectious enthusiasm for this project still brings a smile to my face as I remember his phone call to me not so very long ago.

The work will go on in partnership with Birmingham supported by myself, PGS and I hope many other companies.

Huw Edwards (Petroleum Geo Services)
July 2007

Preface

We know that the seas around Britain contain an immense wealth of archaeological sites and remains, potentially without equal elsewhere in the world in terms of their number and diversity. Despite this our detailed knowledge, necessary to promote effective management, is relatively poor and more often than not based on individual sites or find-spots, lacking the opportunity to take a landscape view. Our coasts and seas are also subject to a seemingly ever-increasing rise in development pressure that represents a risk of damage or destruction to the historic environment, which in itself is unique and irreplaceable.

English Heritage is the statutory advisor to the UK Government on England's historic environment, both on land and within the English Territorial Seas and we are committed to:

- helping people develop their understanding of the historic environment;
- working to get the historic environment on to other people's agenda;
- enabling and promoting sustainable change to England's historic environment;
- assisting local communities to care for their historic environment;
- stimulating and harnessing enthusiasm for England's historic environment, land and sea.

The 3D seismics of the Southern North Sea research programme helps us to further these aims through developing new approaches that rely on more partnerships, strategic engagement, speed and flexibility, and clarity and consistency of advice to industry, commercial awareness and customer service. Furthermore, the outcomes of the research described in this volume are of particular benefit to the aggregate extraction industry, clearly justifying our decision to support the programme through the Marine Aggregates Levy Sustainability Fund, of which we are a Distributing Body on behalf of the UK Government.

The research programme is also timely in that we are involved in the intensive development of a new heritage protection regime, introduced by the UK Secretary of State for Culture, Media and Sport in November 2002, that proposes innovative changes to the types of archaeological site that could be protected by legislation to include the evidence of past occupation, or use, by humankind, at a landscape scale (i.e. area designation) regardless of whether the monument lies on land now, or has been subsequently submerged under the Territorial Seas.

The University of Birmingham research also starkly reminds us that, in relation to the latter point, such remains do not in any way respect present-day administrative boundaries, and the submerged prehistory of the North Sea has value for us all. However, the vast subject area (23,000 square kilometres, yet analysed in 18 months) encompasses jurisdictions from Territorial Seas, and Continental Shelves or Controlled Waters, of many countries, with all the complications that brings in relation to legislative powers, management opportunities to further research, amenity and education, for the benefit of all.

The research has been comprehensive: reviewing a range of available methodologies; appraising the geotechnical cores available for ground-truthing; unlocking previously unknown heritage management and research value from legacy commercial seismic data; developing innovative visualisation techniques; integrating marine geological interpretation; incorporating geophysical data, palaeoecological analysis and dating – all at a landscape scale. In total over 690km of palaeo-coastline was observed, together with the interpretation of 10 major estuaries, and extensive areas of salt-marsh, intertidal zone, over 1600km of fluvial systems and 24 lakes/wetlands.

The implications for heritage management are also considered, acknowledging that the data generated are one of the largest samples of a well-preserved submerged Holocene landscape anywhere in Europe, indicating the potential for survival of submerged Early Mesolithic coastal sites (c. 10,000 – 8500 BP) to supplement our sparse terrestrial record. Information on adaptation to coastal change during the later Mesolithic (c. 8500 – 5500 BP) and the increasing insularity of British prehistory can also be obtained.

Finally, this work is more than just of academic interest. Climate change, global warming and sea level rise are all issues in the forefront of everyone's minds now. So as well as exploring and interpreting, in unparalleled detail, one of the most extensive, yet least known, prehistoric landscapes in Europe, this research will lend context and time depth to present day challenges for us all.

Ian Oxley, Head of Maritime Archaeology, English Heritage
July 2007

x

Acknowledgements

The North Sea Palaeolandscapes Project was a major achievement for all who worked on it but we would like to record here our debt to Dr Kenneth Thomson. Ken was the lecturer in Basin Dynamics at Birmingham, an acknowledged expert on the interpretation and visualisation of seismic data, and a Principal Investigator on the project. Tragically, he died as the project concluded on the 18th of April 2007. To the project staff Ken was a pioneer and an inspiration. To those who knew him beyond the project he remains, in our memories, a great friend and irreplaceable colleague.

Ken, of course, would have recognised that the North Sea Palaeolandscapes Project could not have been attempted without the material and intellectual support of many people and organisations. Our first debt must surely be to PGS Ltd who provided the data used in the original pilot study and the current project study. The kind support provided by Huw Edwards deserves specific mention. The Aggregates Levy Sustainability Fund (managed by English Heritage) provided the financial support without which the project could not have been carried out. Thanks to Kath Buxton, Virginia Dellina-Musgrave and Dr Ingrid Ward for all their help and support over the last two years. Following this, we would acknowledge the work and support of all the members of the project management committee who included; Dr Andrew Bellamy (BMAPA, UMD), Mr Chris Loader (PGS), Dr Virginia Dellino-Mugrave (English Heritage), Mark Dunkley (English Heritage), Mr Huw Edwards (PGS), Dr Joe Holcroft (CEMEX), Dr Justin Dix (University of Southampton), Dr Nic Flemming (University of Southampton), Mr Paul Hatton (Information Services, University of Birmingham), Ms Valerie Scadeng (MCS), Mr Jason Aldridge (MCS), Dr Ingrid Ward (English Heritage).

Professors Geoff Bailey (York) and Martin Bell (Reading) were supportive of the project throughout and assisted greatly in the concluding project seminar when, respectively, they provided the introductory and summary papers.

We would specifically like to thank HP[1] (Dr Martin Walker, Ben Sissons and Nick Hatchard), Mercury Computer Systems[2] (Valerie Scadeng) and Fakespace[3] (Richard Cashmore) for assistance in the launch of this book.

In addition the following provided much needed advice and support:

Mr Graham Tulloch (British Geological Survey), Mr Andy Flaris (BP Exploration Operating Co. Ltd.), Ms Karen Martin (BP Exploration Operating Co. Ltd.), Mr Paul Henni (British Geological Survey), Ms. Lia de Ruyter (Geological Survey of the Netherlands), Dr Ian Selby (Hanson Marine Aggregates), Mr Rob Ingram (Hanson Marine Aggregates), Mr Joe Holcroft (Cemex), Dr Andrew Bellamy (United Marine Dredging Ltd), Dr Ceri James (British Geological Survey), Professor Mike Cowling (The Crown Estate), Dr Mike Howe (British Geological Survey), Dr Colin Graham (British Geological Survey), Miss Claire Pinder (Dorset County Council), Mr Martin Foreman (Hull City Council), Mark Rae (Shell UK Limited), Arthur Credland (Hull City Council), Dr Beryl Lott (Lincolnshire County Council), Eileen Gillespie (British Geological Survey), Rupert Hoare (WesternGeco), Marcia Ritthammer (BP), Mark Bennet (Lincolnshire County Council), Dr Gordon Edge (British Wind Energy Association), Sandra Lane (WesternGeco), Dr Bruno Marsset (IFREMER), Ian Taylor (Boskalis Group), Graham Singleton (CEMEX UK Marine Ltd), Gavin Douglas (Fugro Survey Limited), Ian Wilson (Serica Energy), Dr Chris Pater (English Heritage), Verona Szegedi (ConocoPhillips UK Ltd), Neil Anderton Lynx (Information Systems), Dr Angela Davis (University of Wales, Bangor), John Rowley (British Geological Survey), Carol Thomson (Chevron Upstream Europe), Dr Jim Bennell (University of Wales, Bangor), Mark Martin (Petroleum Geo-Services), Ann Richards (Devon County Council), Georgia Boston (Npower Renewables), Dick Lyons (GEMS-UK), Julie Wallace (Centrica), Matthew Coward (CADW), Katy Whitaker (English Heritage), David Gurney (Norfolk County Council), Serena Cant (English Heritage), Dr Penny Spikins (University of York), Zoe Crutchfield (Joint Nature Conservation Committee), Margaret Stewart (Imperial College, University of London), Professor Greg Tucker (CIRES, University of Colorado, Boulder), Mr Steve Wallis (Dorset County Council), Mr Nick Brown (Norwest Sand and Ballast Co.), Dr Piers Larcombe (CEFAS), John Bingham (TCE/Royal Haskoning UK Ltd), Andy Barwise (Lankelma), Neil Birch (National Wind Power), Brian Ayers (Norfolk County Council), Phil Harrison (DTI), Gwilym Hughes (CADW), Roger Jacobi (British Museum), Malcom Pye (DTI – HSE), Ron Yorston (Tigress), Paul O'Neill (Tigress), Stephen Shorey (Tigress), Dierk Hebbeln (University of Bremen), Tilmann Schwenk (University of Bremen), Hannah Cobb (University of Manchester), Mark Dunkley (English Heritage), Rachel Baines (BGS Digital Licence manager), Joe Bulat (BGS), Paul Henni (BGS), David Long (BGS), Mr Simon Buteux, (University of Birmingham), Professor Geoff Backey (University of Birmingham), Professor Martin Bell (University of Birmingham).

At Birmingham we must record the encouragment provided by the Pro-Vice Chancellor, Professor Geoff Petts, and our respective heads of department, Professor Ken Dowden (IAA) and Professor Paul Smith (GEES). Dr Andy Howard acted

---

[1] http://welcome.hp.com/country/uk/en/welcome.html
[2] http://www.tgs.com/
[3] http://www.fakespacesystems.com/

as academic reader for the publication. Helen Gaffney formatted the text for publication and also assisted in reading the proofs. Henry Buglass worked assiduously to prepare the illustrations for publication and we thank him for the marvellous job he did. Graham Norrie prepared photographic illustrations for which we thank him. Paul Gaffney and Dr Niall McKeown kindly provided translations for the project abstract. The staff at Birmingham Archaeology were unfailingly supportive throughout the project but we would like to thank Caroline Raynor and Alex Jones specifically. Within the VISTA division we would wish to acknowledge the help and support of our colleagues including Dr Henry Chapman, Keith Challis, Mark Kincey, Steve Wilkes and Meg Watters.

1 Mapping Doggerland

Vincent Gaffney and Kenneth Thomson

Eventually, all things merge into one, and a river runs through it. The river was cut by the world's great flood and runs over rocks from the basement of time. On some of the rocks are timeless raindrops. Under the rocks are the words, and some of the words are theirs. I am haunted by waters.
Norman Maclean (1902-90). A River Runs Through It

1.1 Introduction

The inundated prehistoric terrain of the North Sea basin remains one of the most enigmatic archaeological landscapes in northwestern Europe. This region was lost to the sea over a period of c. 11,000 years following the last glacial maximum and the change in relative sea levels resulted in the loss of an area larger than the United Kingdom (Coles 1998). The region therefore contains one of the most extensive and, presumably, best preserved prehistoric landscapes in Europe (Fitch et al. 2007). Moreover, during the Mesolithic, the period primarily covered by this report, the area was probably an important habitat for hunter-gatherer communities (Morrison 1980, 118). This vast archaeological landscape provides Europe with an immense challenge. How are we to investigate, interpret and manage the heritage of this extraordinary, but largely inaccessible, landscape?

This latter point is of prime importance. Although inaccessible and, in most senses, invisible, the archaeology of the region is as fragile as any terrestrial correlate. In terms of mineral and natural wealth the North Sea basin is a strategic resource for the United Kingdom and all the countries that surround it. Its geographical position ensures that this extensive region also functions as a key infrastructural and communications locus (Fleming 2004, 113 - 117). The area is therefore under intensive developmental pressure from a range of threats including mineral extraction and the direct impact of construction. Specific threats range from the laying of pipelines to, more recently, the development of wind farms, the wider issues of mineral extraction and the extensive, generalised, impact of fishing and commercial trawling (Dix et al. 2004, section 1.4). The implication of such threats, in environmental terms, is probably apparent to most aware individuals and organisations with an interest in the region. However, the significance of the southern North Sea is raised in cultural terms when one considers that whilst the continental shelf retains, arguably, the most comprehensive record of the Late Quaternary and Holocene landscapes in Europe (Fitch et al 2005), this landscape was also extensively populated by humans and at specific periods may well have been a core habitat at a European level (Coles 1998; Flemming 2004).

1.2 The context of study

This potential of the southern North Sea for geological and archaeological research was recognised early, by Sir Clement Reid, in a book on the submerged forests of the United Kingdom published in 1913. Here Reid noted, in a remarkably perceptive paragraph that *"the geologist should be able to study ancient changes of sea-level, under such favourable conditions as to leave no doubt as to the reality and exact amount of these changes. The antiquary should find the remains of ancient races of man, sealed up with his weapons and tools. Here he will be troubled by no complications from rifled tombs, burials in older graves, false inscriptions, or accidental mixture. He ought to here find also implements of wood, basketwork, or objects in leather, such as are so rarely preserved in deposits above the water-level."* (Reid 1913, 9).

Following this promising start, the pioneering work of Sir Harry Godwin on moorlog (peat) deposits associated with the 1931 Colinda harpoon find from the Leman and Ower banks, demonstrated the capacity of these extensive submerged deposits, to provide paleoenvironmental evidence and proved their terrestrial origin, (Burkitt 1932, Godwin and Godwin 1933). Shortly after, Sir Graham Clarke's (1936) seminal work on the "Mesolithic Settlement of Europe" acknowledged the probable settlement potential and the cultural significance of the area. It is notable, however, that these early initiatives were not substantively built upon. Whilst this must have largely been a consequence of inaccessibility of the archaeological deposits Clement Reid (1913, 3) also, presciently, predicted that *"the archaeologist is inclined to say that* [these deposits] *belong to the province of geology, and the geologist remarks that they are too modern to be worth his attention; and both pass on."* The demise of active archaeological research across the North Sea basin from the mid twentieth century was paralleled by the marginalisation of the presumed archaeology of the area. Whilst not denying that some archaeologists were aware of the archaeological potential of the region, the area was increasingly interpreted or represented as a land bridge from mainland Europe to Britain (Coles 1998). The largely unspoken implication was that the inundated area was unimportant in cultural terms (Coles 1999, 51). In many ways it might be said that there was a spiral of indifference towards the archaeology of the region.

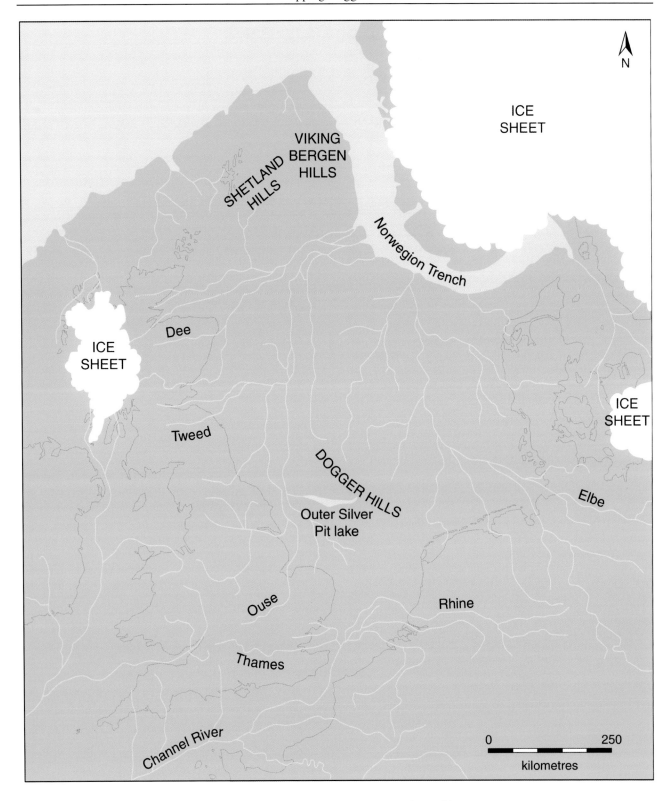

Figure 1.1 Hypothetical maximum extent of Doggerland (redrawn from Coles 1998)

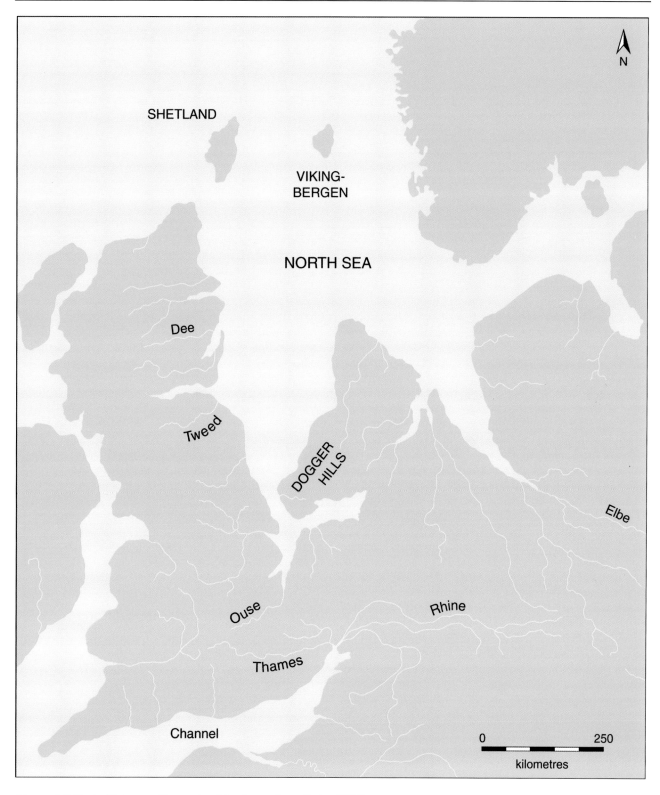

Figure 1.2 Early Holocene Doggerland (redrawn from Coles 1998)

More recently, the significance and potential of the archaeological and geomorphological record of the southern North Sea has become an emerging academic interest (Flemming 2004). Within archaeology this phenomenon can be traced directly to the 1998 review article by Professor Bryony Coles, "Doggerland: a speculative survey" (Figure 1.1 and Figure 1.2). After Coles, Dr Nic Flemming has worked unceasingly on promoting the archaeology of the area, most notably through his recent edited volume on the archaeology of the region (Flemming 2004). The ALSF-funded assessment of the archaeological potential of the British continental shelf by Dix et al. (2000) also provided a significant context for further research around the coast of Britain.

However, the fundamental consequence of these publications has actually been an increasing awareness of the deficiencies of our knowledge of the North Sea in terms of the nature or extent of the archaeological deposits of the region. Such sentiments have, more recently, been echoed in the contents of a series of Department of Trade and Industry regional strategic environmental assessments, or "SEA" volumes, for the mainland British marine territories (Flemming 2002, 2003, 2004b, 2005; Wickham-Jones and Dawson 2006).

The lack of knowledge associated with the North Sea Holocene surfaces was so profound that, as recently as 2004, Flemming noted that the inundated landscapes of the Southern North Sea were essentially *terra incognita*. This profound lack of knowledge was maintained despite the results of geological studies that suggested that sediment in the area that might be associated with human occupation achieved depths of 1 to 5m thick and, locally, a maximum thickness of 30-40m (Laraminie, 1989). The potential of these substantial, unexplored deposits has been underscored by the significant number of human artefacts and mammal remains that are often trawled or dredged from the region (e.g. Van Kolfschoten T. and Van Essen 2004). It is usually assumed that such finds originated from eroding or disturbed seabed deposits (Flemming, 2002; Glimmerveen et al. 2004). Flemming (2002) suggested that richer environments for the origin and preservation of archaeological materials could include Holocene fluvial valleys and the Outer Silver Pit, a vast sea inlet which existed to the south of the Dogger Bank from 8,000 - 7,500BP.

These general impressions were supported by the increasing density of sites located around contemporary coasts that, presumably, can be extrapolated onto inundated coastlines beneath the North Sea. This information, clearly suggests that the lack of material associated with deeper waters indicates an absence of evidence rather than evidence of absence (Fischer 2004, figure 3.3; Pedersen et al 1997). The paradox of the North Sea, therefore, is that although the environmental and cultural potential of the region remains largely unknown, it may still be correct to suggest the landscape archaeology of the region is significant at a global level (Mithen 2003, 154-157). Sourcing inundated deposits, and thereby providing an option to protect surviving archaeology, is a key, but problematic goal.

1.3 Previous methodological approaches

If our knowledge of the archaeological deposits of the North Sea is so tenuous, it might be hoped that the larger geomorphological context of the region offers the opportunity to make general observations on the potential nature of preserved archaeological deposits. Unfortunately, although the North Sea has been the subject of extensive exploration for a variety of commercial or academic reasons for decades, this is probably not the case. Our current understanding of the morphology of the Holocene landscape of the southern North Sea is largely based on bathymetric data. This is supported by considerable exploratory activity by the geological services of countries bounding the sea and commercial groups seeking to exploit the area. Work by Jelgersma (1979) produced a series of highly influential maps for the major changes in the coastline from 18,000BP to 8,300BP and, significantly, noted the formation of an island at the Dogger Bank around 8700BP. An attempt was then made to place this landscape within a cultural context by Coles (1998) who dubbed the emergent plain "Doggerland". This work contained hypothetical reconstructions of the coastline from the Weichselian maximum through to 7000BP, but was ultimately based on the earlier study by Jelgersma. Whilst this approach has provided an overview to the area it remains true that the palaeogeography of the region remained lacking in critical detail. Researchers, including Lambeck (1995), Shennan (2000), Shennan and Horton (2002) and Peltier (2004), have used isostatic rebound models to help constrain and improve the present bathymetry-based models. This has resulted in minor modifications to current coastal models but the lack of detail within the landscape (e.g. the location of fluvial systems, details of coastline etc), and the failure to incorporate late Holocene and recent sedimentation, still remain significant issues (Bell et al 2006, Box 1, 14). In so far as these factors have the effect of masking the true relief of the palaeo-landscape it is unlikely that an adequate appreciation of the human landscape can be achieved using data provided by previous studies.

In methodological terms, therefore, the investigation of past marine environments has generally been limited by available data that had serious limitations. These have included:

1. Seabed sampling and shallow coring: These provide high quality chronological, sedimentological and environmental data. However, data is widely spaced and provides a poor spatial framework and thus limits its use in assessing the larger landscape and its archaeological significance or potential.
2. High resolution 2D seismic: Traditional shallow seismic techniques (e.g. Stright 1986; Velegrakis et al., 1999) have provided detailed information on the architecture of sedimentary systems but as the data is generally acquired as a series of 2D profiles, a weak three-dimensional framework is created due to the necessary interpolation between the profiles.
3. High resolution 3D seismic: These data represent a significant advance in imaging shallow geology (Bull et al 2005, Gutowski et al. 2005, Muller et al. 2006), but the centimetre-scale resolution of the data dictates that only small areas ($<1km^2$) can be realistically surveyed.
4. High resolution bathymetry: This may provide excellent images of the seabed topography and is

capable of providing detailed images of Late Pleistocene and Holocene features that have a bathymetric expression. Whilst bathymetry provides a reasonable approximation for the land surface for the area it can rarely consider, or attempt to resolve, burial of features that may have occurred during or after submersion (Cameron et al., 1992). Consequently, the technique is unsuitable for areas including the southern North Sea and the Irish Sea, where deposition has buried most of the Quaternary and Holocene. The scale of this problem was clearly stated by Dix et al. (2004, 89); *"although modern bathymetry can correlate to surfaces relating to earlier periods, in many instance there may be a significant difference (up to c. > 20 m) between them. This can lead to inaccurate representations of shoreline positions (up to 60 km difference) and past topography can be markedly misinterpreted. The bedrock horizon represents a minimum value that could be used in reconstruction. However, modern bathymetry does not represent a maximum value as processes of erosion may have reduced its height over time"*.

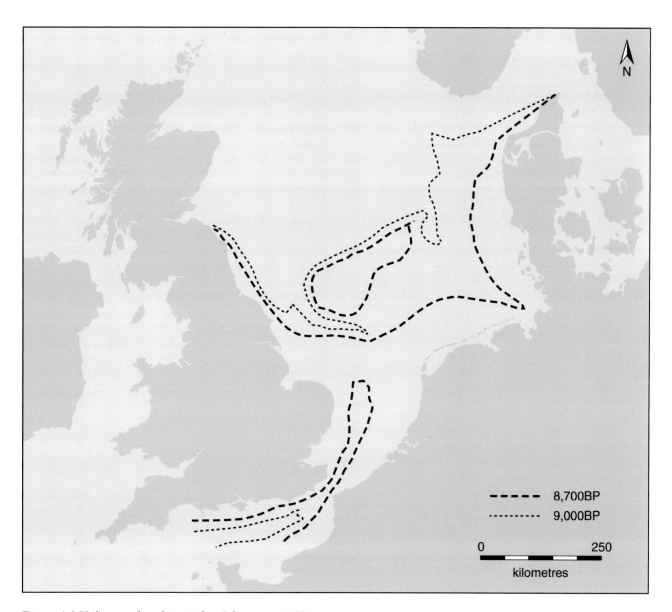

Figure 1.3 Holocene shorelines (after Jelgersma, 1979)

The limitations of these methodologies are also apparent in the archaeological literature. There is considerable interest in the investigation of marine features and the identification of marine landscapes. However, the available technology and scale of archaeological application has tended to restrict studies to the immediate coastal zone and to relatively small, intensively surveyed areas (Mueller et al 2006). Whilst this has been adequate for exploration of known sites (usually of the historic period, e.g. Paoletti et al. 2005) or micro-regional survey (Pedersen et al 1997), it has largely precluded major landscape exploration. Prior to the current work, therefore, there was no plausible topographic or geomorphological context that could provide a credible proxy indicator for human activity across the former Holocene landscapes of the North Sea.

1.4 Towards an alternative methodology

The impetus and opportunity to develop a methodology using 3D seismic data to deal with this challenging landscape derived from doctoral research carried out at the University of Birmingham by Simon Fitch and under the supervision of the project's principal investigators, Gaffney and Thomson (Fitch et al. 2005). The 3D seismic datasets acquired on the United Kingdom continental shelf for exploring deep geology represent a major resource for understanding Late Pleistocene and Holocene geology. With extensive regional coverage and spatial resolutions of c. 12.5m such datasets provide the opportunity of mapping relatively recent geology at a regional scale and with relative speed. Standard geophysical interpretation techniques usually used on such data to explore deeper features, augmented by volume and opacity rendering, provide significant advantages in reconstructing palaeogeographies and allow the true 3D architecture of Late Pleistocene and Holocene systems to be established (see Thomson and Gaffney, this volume).

The original research at Birmingham coincided happily with an emerging requirement to manage the archaeological heritage in the light of aggregate extraction within the area. Funding for a larger project was made available to the Birmingham team through the Aggregates Levy Sustainability Fund. This fund, administered by English Heritage, seeks to promote best practice in planning aggregate extraction and to provide data to support the protection of our marine heritage that may be impacted by such activities[1]. This serendipitous opportunity permitted the team to develop a methodology centred around the use of extensive 3D seismic data to map Holocene features across a large area of the southern North Sea. A team of three researchers was initially employed to work on this data; Kate Briggs, Simon Fitch and Dr Simon Holford. The papers presented in this volume present the results of this work.

The surfaces investigated as part of this project effectively represents the Holocene landscape inundated between 10,000 and 7,500BP and, in archaeological terms, are associated with the Mesolithic period (Cameron et al. 1992; Jelgersma, 1979; Lambeck, 1995). Given the origin of the data the study area was defined by the extent of available data rather than the probable historic Holocene shorelines (Figure 1.3) or notional areas defined for other purposes (see Cameron et al 1992 for the BGS definition of the Southern North Sea region).

Data for the Southern North Sea is provided through a research agreement between the University of Birmingham and Petroleum Geo-Services[2] (PGS) and we are particularly indebted to Mr Huw Edwards for facilitating access to this information. PGS MegaSurveys are based on seismic data that have been released by oil companies, PGS owned seismic surveys and non-exclusive seismic data made available through other geophysical contractors. Usually these data are available as 3D time migrated seismic surveys. Although quality controlled the different age and data acquisition methods used to collect data demand that the seismics vary in quality[3].

Figure 1.4 illustrates that these data exist as a significant continuous data source across much of the Southern North Sea. Whilst the data does not currently stretch coast-to-coast the total full-fold area of coverage of the Southern North Sea Megasurvey is in excess of 23,000 km² and represents more than 60 original 3D surveys belonging to 20 different data owners. Altogether, this data set represents the largest available data source for the exploration of the palaeogeography of the Southern North Sea region and, in archaeological terms, constitutes the largest contiguous archaeo-geophysical survey programme ever attempted. The work also follows the tradition of seismic study and large-scale archaeological remote sensing projects managed at Birmingham (Gaffney et al. 2000, Thomson 2004, Barratt et al. 2007).

Within this context, the specific aims of the project were:

- To use the existing 3D seismic datasets acquired on the United Kingdom continental shelf for exploring Late Quaternary and Holocene geology over an area of the Southern North Sea.
- To provide maps of the recent geological sequence at a regional scale.
- To provide detailed digital mapping of the topographic features of the region and to use voxel rendering to allow the true 3D architecture of Late Quaternary and Holocene systems to be established.
- To compare the Holocene topographic data with available core and borehole data to ground truth data and calibrate results.
- To provide a model of survival potential for environmental and archaeological deposits within the area of the Southern North Sea to be used by the aggregates industry to plan extraction and mitigation strategies.
- To use data on environmental and archaeological potential to provide an extensive depositional

---

[1] http://www.english-heritage.org.uk/server/show/nav.1315

[2] : http://www.pgs.com/)

[3] http://www.pgs.com/business/geophysical/research/library/mc3d/dbaFile7567.html?1=1&print=true

map of the Southern North Sea for use for aggregate developmental purposes.
- To utilise seismic attribute analysis to map depositional systems in detail and to make calibrated lithological predictions that may be used in aggregate deposit modelling.
- To provide palaeocoastline data, which may be used in the development and calibration of current sea level and palaeobathymetry models.
- To disseminate knowledge of the methodology and outcomes of the project for the purposes of supporting and developing the aggregate industry and management of the mineral resource.

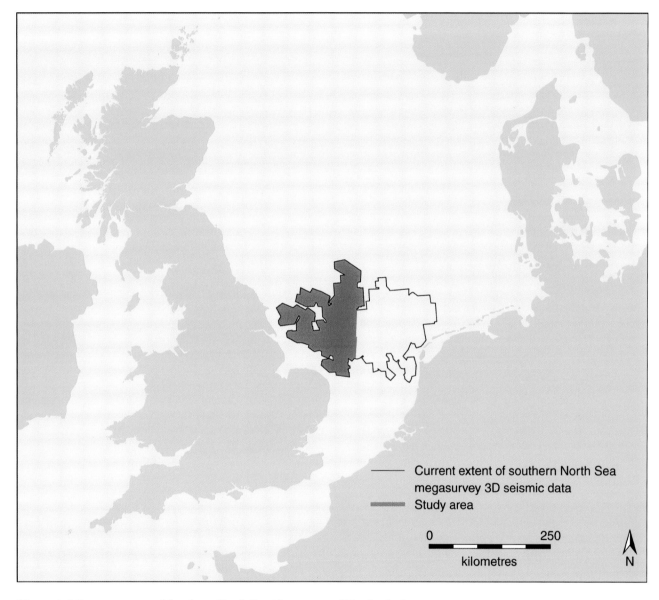

Figure 1.4 Current extent of Southern North Sea Megasurvey 3D seismic data (source PGS http://www.pgs.com/business/products/datalibrary/nweurope/southernnorthsea/snsmegasurvey/)

A full description of the technologies utilised to explore and integrate the available seismic data is provided in the papers by Thomson and Gaffney and Fitch et al. (the atlas, this volume). It is enough to note here that the significance of the first results of this work was rapidly appreciated, and that mapping of the area proceeded apace, as is demonstrated in the atlas paper by Fitch et al. This is complemented by papers from Briggs et al. and Holford et al. (this volume) that demonstrate the detail of specific geomorphological structures, including the nature of internal features identified within the Outer Silver Pit.

It is true, however, that as the project proceeded there was an increasing concern within the team concerning not so much the extent of available supporting data (including 2D seismics lines, cores etc) but the quality, or even availability, of some of this information. In particular, spot-checking of cores for environmental potential suggested that the description of core data was, in some cases, misleading and that the environmental potential of some samples might have been compromised by their storage (Smith et al. this volume). Consequently, a project variation was submitted to English Heritage that supported a data audit to assess the extent of available 3D coverages, the availability of other supporting datasets and their potential for research. Dr Mark Bunch was employed for these purposes and the results of this important work are presented in summary within this volume.

In providing this extended introduction it is worth considering the wider significance of this study. Initially, it should be stressed that this volume represents our initial, tentative steps towards providing a robust methodology for the investigation of deeply buried and inundated historic land surfaces. The results of the North Sea Palaeolandscapes Project are, we firmly believe, a major contribution to our understanding of the Holocene land surfaces of the North Sea. From a methodological perspective this is of enormous significance. However, the Holocene landscapes discussed here do not represent the total of available data for the British continental shelf. Comparable areas of submerged, but previously habitable landscapes, can also be found in the Black Sea (Ryan and Pitman 2000; Ballard et al. 2000), the Florida Gulf (Stright 1986; Faught 1988; Marks and Faught 2003; Faught 2004), the Gulf of Arabia (Lambeck 1996) and a number of other regions of the world (e.g. Dortch 1997; Bailey 2004), many of which have also been subject to extensive exploration for mineral extraction. The work presented here is therefore replicable elsewhere and, if implemented, the results for regional research are likely to be as exciting and challenging as those derived for the Southern North Sea.

Of course, there is room for development. This project, which lasted for a mere 18 months, would have benefited from a more substantive integration of supporting information, including high resolution 2D seismic data and further core data. Unfortunately, the audit carried out as part of the study suggests that existing data will not always be available or sufficient for the purposes of refinement or ground truthing of results. There is, therefore, a real need for dedicated, expensive ship time to provide new data to ground truth and extend the results of this study.

Despite these observations, the scale of the work and the fact that the landscape transcends national boundaries ensures that, aside from primary archaeological or geomorphological output, the implications of the results are of international significance in terms of heritage management, at the very least. We have presumed, for nearly a century, that the North Sea contained a significant archaeological record but it has always been a challenge to manage a resource that was largely inaccessible, entirely unpredictable and, essentially, a hypothetical construct. The results presented here suggest that this record may be traced, in part, through the reconstruction of the topographic context of the region. As a consequence, the heritage agencies of countries bounding the North Sea may well have to re-assess their marine management strategies in the light of this information. In this context the steps toward a historic landscape characterisation methodology, as described in the final paper of the volume, are we believe an important contribution towards the management of problematic, marine landscapes.

Ultimately, the principal achievement of the project has been to explore and begin to interpret in unparalleled detail one of the most extensive, yet least known, prehistoric landscapes in Europe. Whilst our knowledge remains imperfect the area is no longer the "terra incognita" pondered upon by Flemming less than 3 years ago (Flemming 2004). Indeed, in the light of our previous lack of knowledge, the scale of the work carried out by this project is truly startling. The analysis of 23,000 square kilometres of seismic data is comparable to carrying out a geophysical survey over a country the size of Wales. It is a cliché to assert that the past is a foreign country. However, in the case of the North Sea Palaeolandscapes Project, it is hardly hyperbole to assert that, along with the outstanding contributions of Coles, Flemming, Dix and others, the project has effectively begun to provide the archaeological outline of a previously undiscovered European realm.

The final point to be made is more emotive. The loss of extensive late Pleistocene and Holocene landscapes, after the last glacial, represents the only previous period during which modern man experienced the impacts of global warming at a scale predicted for the next century. The North Sea Palaeolandscapes Project provides quantitative and visual evidence for the nature and significance of such change. The recreation of the Mesolithic landscape and coastline may, ultimately, be factored into improved coastal models and this is a practical and desirable outcome. We should not forget, however, that this was a populated land. The loss of such extensive areas, insidious and slow overall but terrifyingly fast at times, must have been devastating for the Mesolithic populations of the great northern plains. The coastlines, rivers, marshlands and hills mapped during this project were, for thousand of years, parts of a familiar landscape to the hunter-gatherers of northwestern Europe. The land and its features would have been named; some areas might have been revered and held personal associations or ancestral memories dear to these peoples. It is almost impossible for us now to comprehend the demise of environments and ecologies that supported communities, tribes and entire peoples. Whole territories may have disappeared within the memory of a single generation, and the stress to the indigenous populations is beyond our experience (Mithen 2003). The memories and associations of cultures disappeared, with the landscape itself, as sea levels rose and the land retreated.

As this project concludes, the UN Intergovernmental Panel on Climate Change is finalising its report on the nature, scale and implication of global warming (http://www.ipcc.ch/). At such a time, and when climate change, global warming and sea level rise are now accepted as amongst the greatest threat to our lifestyles, the fate of the Holocene landscapes and peoples of the North Sea may yet be interpreted, not as an academic curiosity, but a significant warning for our future.

2 Coordinating Marine Survey Data Sources

Mark Bunch, Vincent Gaffney and Kenneth Thomson

2.1 Introduction

The North Sea Palaeolandscapes Project (NSPP) has primarily relied upon the large Southern North Sea (SNS) 3D seismic MegaSurvey developed by Petroleum GeoServices (PGS). This is a regional merge of surveys acquired by the petroleum industry, oil and service companies over the last 20 years (Terrell et al., 2005). It offers an unparalleled source of data for visualising and interpreting buried features of the emergent Holocene landscape at a regional scale providing a broad spatial framework to explore the offshore prehistoric landscape following the last glaciation.

As stated by Thomson and Gaffney (this volume), 3D seismic survey data is one of a diverse suite of survey data types that exists within the SNS. Different surveys provide varying information about the seabed and subsurface. Survey design dictates the penetration through the subsurface as well as the spatial resolution of the data set. This clearly has implications for the survey scale. Owing to the location of viable hydrocarbon resources and the logistics of development, the SNS MegaSurvey is confined to a region lying at least 11 miles off the English coast. There is a so-called "white band" between the coast and the MegaSurvey that that may be crossed only by smaller scale surveys or 2D seismic lines. If we are to understand how prehistoric societies interacted with the evolving postglacial shoreline, an appropriate strategy must be developed to utilise surveys conducted within the white band. In addition, we will gain greater insight into behavioural patterns in the hinterland by augmenting our broad model of the emergent landscape with more focussed interpretations using survey data at a finer scale. This paper seeks to investigate the variety of survey data available to serve these purposes and also to assess their relevance to the broader aims of the North Sea Palaeolandscapes Project (NSPP) as identified in Gaffney and Thomson (this volume).

2.2 Identification of sources: who acquires, owns or holds survey data?

There are three groups that require offshore survey data: governmental organisations, industry and academia. Governmental organisations, including conservation and regulatory bodies, survey for monitoring and engineering purposes as dictated by legislation and infrastructural planning. For example, former public service organisations including British Telecom, are responsible for regularly surveying seabed cable and pipeline routes to monitor their structural integrity. The British Hydrographic Office continuously maps the sea floor at a fine spatial resolution to ensure the latest navigation charts are accurate. Conservation and regulatory bodies also undertake surveys to monitor environmental change arising from seabed use.

Current government legislation promotes renewal and investment in offshore industrial sectors. This is particularly notable in respect of the growth in renewable energy projects. Offshore 'green' energy generation is seen by the government as a key strategic objective in the battle to mitigate anthropogenic climate change. Funding systems including the Aggregates Levy Sustainability Fund (ALSF) have involved academia directly in combining research and development with monitoring and evaluation.

Offshore industrial surveys are specifically designed with industrial application in mind. Oil companies carry out large-scale acoustic surveys to visualise geological structures and stratigraphy at depth. Smaller-scale, shallow geotechnical surveys are carried out on their behalf for engineering and compliance purposes. Other industries, including marine aggregates producers, specifically target the shallow marine sediments as a resource. These groups regularly conduct 2D seismic sub-bottom profiling, high resolution, seabed bathymetric mapping using sonar and direct sediment sampling by shallow core or grab. Another growing industrial application within territorial and offshore UK waters is the development of large offshore wind farms. Aside from carrying out surveys in order to estimate their direct environmental impact, wind farm proposals require engineering survey to ensure the safe installation and anchor of turbines to the seabed.

The activity of the British Geological Survey (BGS) uniquely occupies parts of all three sectors. The BGS is a public sector organisation that is a component of the Natural Environment Research Council (NERC), a body reporting to the Department of Trade and Industry (DTI). Significantly, it is also an academic body undertaking applied research that is part-funded by the government and its own commercial activities. The BGS has undertaken extensive survey of the shallow seabed sediments in UK territorial and offshore waters over many decades.

2.3 Key data repositories

Under its remit for strategic research and surveying, the BGS has compiled the national archive of marine shelf seabed sedimentary and geological maps. These are derived from shallow geophysical survey and sediment sampling. Figure 2.1 illustrates the distribution of acquired shallow geophysical survey lines and shallow sediment samples around the UK according to the BGS data archive web GIS, GeoIndex, as of October 2006. It is clear that geophysical surveying is possible within all waters beyond a certain depth. Direct sediment sampling was achieved closer to shore and is particularly concentrated within enclosed estuaries and embayments.

A variety of geophysical survey techniques have been utilised as part of this process. In order to map the shape of the sea floor all areas have been explored using an echo sounder and some by more sophisticated sonar techniques. The majority have been surveyed using a potential field instrument: a gravimeter or magnetometer. These give an indication of a variation in bulk density and magnetisation of seabed material laterally. However, the most useful technique for visualising the vertical stratigraphy of seabed sediments is sub-bottom seismic profiling. Most of the BGS geophysical survey lines have been covered by a seismic reflection survey using either a boomer, sparker or pinger acoustic source (Delgado 1998, Gillespie pers. comm. 2007). Each of these is an industry name that refers to the method for generating the acoustic pulse that propagates radially outward, through the water column and underlying sediment, to be reflected back at interfaces of contrasting density. BGS 2D seismic survey data is particularly suitable for mapping of detailed features as these were acquired using a high-resolution single channel receiver (hydrophone) system. The boomer generates a source wavelet that tends to produce the clearest results for interpreting shallow features.

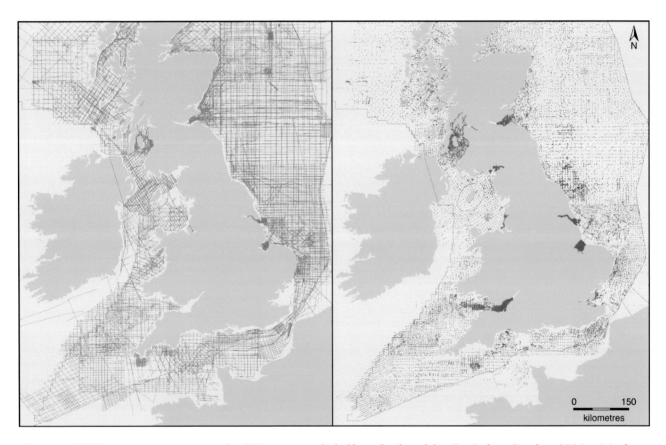

Figure 2.1 BGS survey coverage on the UK continental shelf as displayed by GeoIndex, October 2006. (a) shows geophysical profile lines, (b) shows shallow sediment sample stations

Sediment samples may be disposed of following cataloguing or may be retained in archive by the BGS. The store, at Loanhead, in Edinburgh, forms part of the National Geosciences Data Centre (NGDC), which offers an archiving and curatorial service for voluntary donations of survey data, samples and field notes by both industry and amateur geoscientists. The NGDC has a section dedicated to conventional prospecting data – seismic surveys and borehole logs – that is divided between onshore and offshore activities. These sections are the UK Onshore Geophysics Library (UKOGL) and the National Hydrocarbons Data Archive (NHDA) respectively. The NHDA can act to relieve UKCS licensee operators of the expensive legal obligation to retain certain survey data in perpetuity for recall by the DTI at any time. However, this obligation does not apply to core samples or cuttings. The decision to donate these rather than dispose of them, is made at the discretion of the operator.

The NHDA is a joint venture between the Department for Trade and Industry (DTI) and Common Data Access Limited (CDA), a not-for-profit subsidiary of the UK Offshore Operators Association (UKOOA), which administers data exchange amongst its members. The BGS manages the NHDA and runs the associated UK Digital Energy Atlas and Library (UK DEAL) project: a public web site aiming to promote data and information relevant to the exploration and production of hydrocarbons on the UKCS. UK DEAL aims to provide a publicly accessible national catalogue of UKCS geosciences data, their sources and owners, to facilitate data sharing between the public and/or prospective licence applicants and data owners/current licensees.

Figure 2.2 shows the distribution of previously acquired UKOOA 3D seismic survey blocks. 3D seismic has been the bulk survey data type used for mapping the broad features of the emergent landscape in the SNS. 3D surveys are ideal for reconnaissance mapping because they are acquired over large areas but can still be manipulated visually to resolve sufficient detail at the scales of interest relevant to this project. Unfortunately, as Figure 2.2 demonstrates, the distribution of these data is uneven and large areas remain without coverage. The large gap off the northeast coast of England is a prime example, and is in an area of great interest to archaeologists (Waddington 2006).

The DTI Guidelines for the Release of Proprietary Seismic Data UKCS 2004, and the Agreement between the International Association of Geophysical Contractors and the DTI for the release of speculative seismic data 2002, govern the availability of survey data featured on the UK DEAL web GIS.

Petroleum industry well site surveys are an excellent source of high vertical resolution 2D seismic survey data. Unfortunately, these data are not 'tied' to the licensed oil and gas licence block and data owners are not required to retain these in perpetuity. Offshore operators and the DTI continue to debate whether digital site survey data should be submitted by law to the NHDA. Currently, only c. 1000 site survey reports have been donated to the NHDA voluntarily. These represent a fraction of the data acquired offshore.

2.4 Governance and surveying within territorial waters

The Crown Estate (TCE) owns the seabed within the 12-mile territorial limit and is responsible for issuing leases and licenses for its commercial and infrastructural development. These include leases of easement for seabed cables and pipelines, licenses for the extraction of minerals, excluding oil, gas and coal (principally marine aggregates), and leases and licenses for the construction of offshore wind farms. Under the terms of *The Marine Works (Environmental Impact Assessment) Regulations 2007* consultation document released by DEFRA in 2006, the regulators of development and disposal activities within territorial and offshore waters of England and Wales are the bodies responsible for licensing. These are the Secretary of State for projects within English waters, and the National Assembly for Wales within Welsh waters. The regulator determines the necessity of conducting an environmental impact assessment (EIA). The environmental and cultural heritage of the seabed within territorial waters is of concern to TCE. However, the ODPM awards licenses and acts as the environmental regulator under advice from conservation bodies including English Heritage (EH).

Marine aggregates comprise up to a quarter of the total sand and gravel used in Britain (Bellamy, 1998). The marine aggregates industry generates almost half TCE's annual marine income and constitutes approximately a third of commercial activities in its marine portfolio (TCE, 2006). Marine aggregates producers generally target shallow glacial sand and gravel banks, usually within the 12-mile territorial limit. Wenban-Smith (2002) emphasises the requirement to work with marine aggregates producers to develop a system for identifying culturally significant deposits in order to identify significant threats prior to disturbance. In particular, he predicts the likelihood that thin, well preserved sedimentary layers with high archaeological potential directly overly Marine Aggregates Deposits (MADs) within the English Channel and SNS. Clearly, the work of recent ALSF projects, in collaboration with Aggregate groups as has been the case with the NSPP, will clarify the extent that this cooperation occurs or can occur given the use of available data for cultural resource management purposes.

In order to gain a licence for aggregate extraction, a marine aggregates group must complete a two-stage application process to TCE and the ODPM. Consultation on a statutory framework is currently underway. At present, this involves resource assessment and an EIA that must address the potential impacts of removing seabed material on coastal erosion, fishing, archaeology and marine life. Both necessarily require geophysical surveying and direct sediment sampling.

Offshore wind farms currently generate almost 16% of the total electrical power generated by UK wind farms. Most offshore wind farms occur within the 12-mile nautical limit and therefore fall under the jurisdiction of TCE for appraisal and licensing. Licensing for wind farm development begins with a request from the DTI to TCE to announce a competitive tender process for new wind farm developments. After pre-qualification, developers must win statutory consents from two government departments, the Offshore Renewables Consents Unit (ORCU), and the Marine Consents Environment Unit (MCEU). This stage involves submission of an EIA that includes consideration of submerged archaeology. After securing these consents a lease or licence is granted by TCE.

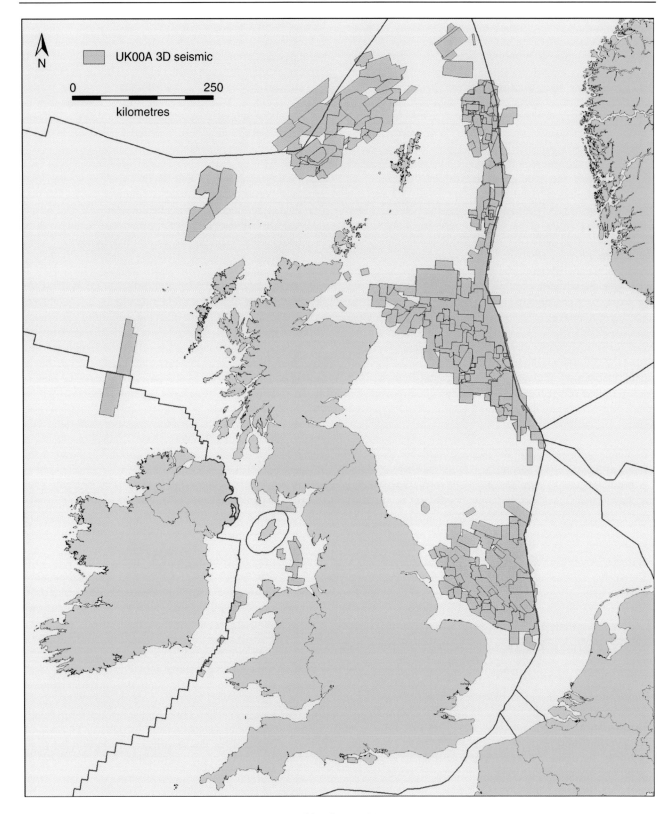

Figure 2.2 Distribution of 3D seismic surveys acquired by the UKOOA

2.5 Public metadata resources

There are a number of web-based GISs detailing available survey data across the UK coastal shelf. The MAGIC GIS (http://www.magic.gov.uk) brings together information on environmental schemes and designations from six UK government organisations: DEFRA; English Heritage (EH); NE; Environment Agency (EA); the Forestry Commission (FC); and the Department for Communities and Local Government. The MAGIC GIS features a Coastal and Marine Resource Atlas theme that presents metadata from offshore regulatory bodies including the DTI and the BGS. These can be queried elsewhere on the project website.

The Mapping European Seabed Habitats (MESH – http://www.searchmesh.net/default.aspx) is an international marine habitat-mapping programme involving a consortium of twelve partners across the UK, Ireland, the Netherlands, Belgium and France. The aim is to produce standardised seabed habitat maps for northwest Europe. The MESH website hosts a basic GIS. A much larger metadata archive of all forms of marine surveying within specified regions can also be queried elsewhere on the website.

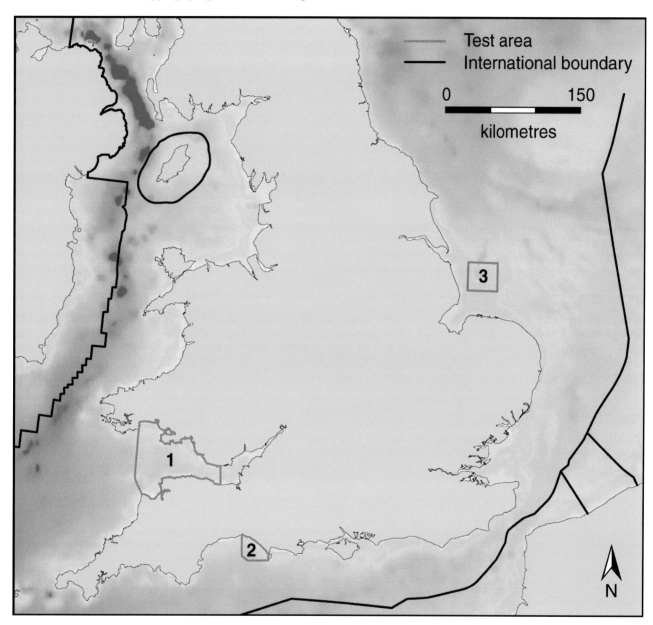

Figure 2.3 Three study areas around the English coast

The British Oceanographic Data Centre (BODC) website (http://www.bodc.ac.uk) hosts the European Directory of Marine Environmental Datasets (EDMED). Here, there is information available for bathymetric, seismic and sediment sampling (core, dredge, grab) surveys undertaken in UK offshore waters by a range of research groups and industrial organisations. These fall under the "Geology – Geophysics – Sedimentation" theme. The BODC is part of the NERC MetaData Gateway, which also provides access to catalogues of data held by the other NERC data centres: British Atmospheric Data Centre (BADC); British Geological Survey (BGS); NERC Earth Observation Data Centre (NEODC); Antarctic Environmental Data Centre (AEDC); Environmental Information Centre (EIC). The NERC MetaData Gateway is in the process of being replaced by a service based on the NERC DataGrid, a web-based search facility that aims *"to make the connection between data held in managed archives and data held by individual research groups so that the same tools can compare and manipulate data from both sources"* (http://ndg.badc.rl.ac.uk/).

The DTI runs a website dedicated to the regulation of the oil and gas industry, and offshore development in general (http://www.og.dti.gov.uk). Users can access a repository of existing development licences and all petroleum industry survey applications (PON14) and decisions since 2001 (*Conservation of Habitats Regulations 2001*). Spreadsheets of metadata associated with these applications identify the applicant company, and the type, time and location of their proposed surveying.

SeaZone Solutions Limited provides a commercial source of marine environment and coastal zone metadata to the crown, the government, industry and academia. They sell a range of GIS data layers relating to near-coastal land, backshore, shoreline and near-shore environments.

Archaeological artefacts recovered offshore constitute archaeological 'ground truthing' of our palaeolandscape models, with due acknowledgement of the frequent lack of an accurate provenance for such finds or indeed the accuracy of associated positional data. There are two sources of such records. Historic Environment Records (HERs) are maintained by local authorities. The National Monuments Record (NMR) is maintained by EH. Only a few coastal HERs incorporate a marine section, and given the lack of secure provenance for many artefacts this must constitute a weakness. EH hope that all will feature a marine aspect eventually (Roberts and Trow, 2002). The marine HERs and NMR will form an essential tool in the management and protection of the marine archaeological resource (Robert and Trow, 2002). To this end, EH has collaborated with 'legitimate users of the sea' to draft the *Protocol for reporting finds of archaeological interest* (2005). This protocol should ensure that the location of artefacts dredged or caught up in fishing nets is recorded as part of the cataloguing process.

2.6 Case study assessments of survey data

As part of the project data audit, which forms the basis of this paper, three areas adjacent to the English coastline were chosen for a pilot study to assess useful survey data that have been acquired within them (Figure 2.3). These areas are geographically distinct and offer different appeals to offshore developers. Two are adjacent to areas containing significant archaeological sites, the other lies just beyond the fringe of the area mapped by the NSPP.

2.6.1 Area 1: the Bristol Channel

The Bristol Channel has the second largest tidal range in the world, 15m during spring tides. This drives a dynamic sediment transport system capable of filling artificial scour holes within a few tides (Murray, 1987; Newell et al., 1998). Area 1 is traversed by a number of pipelines and international telecommunications cables that continue west to Europe and beyond the Western Approaches. These must be surveyed annually to assess their structural integrity. Area 1 also supports three licenses to dredge marine aggregates and a single wind farm licence. Figure 2.4a summarises these known threats to seabed archaeology.

BGS survey here has been dominated by sediment sampling, though the channel is relatively well covered by 2D seismic profiles shot by both the BGS and the petroleum industry (Figure 2.4b). BGS survey data is straightforward to obtain by purchase order at standard rates. Unfortunately, the cost of BGS data may well be prohibitive to extensive or academic exploration and this may yet prove a major barrier to promoting a marine CRM agenda. Most of the petroleum industry 2D seismic survey data shown in Figure 2.4 belong to BP Exploration. Petroleum companies are often amenable to cooperating with academic requests for study in the spirit of the *DTI Guidelines for the Release of Proprietary Seismic Data UKCS, 2004*. Individual data release agreements must be drafted, signed and countersigned by both parties. These data are relatively old (early 1970s to early 1980s) and so, in the absence of active exploration within the Bristol Channel, may be readily available.

The BGS has recently been involved in an Aggregates Levy Sustainability Fund (ALSF) project investigating seabed ecology in the Outer Bristol Channel. Part of the project involved the acquisition of 2D seismic sub-bottom profiles (Figure 2.4d). It is thought that these profile data have not been interpreted with respect to seabed archaeology. They would provide an ideal dataset traversing prehistoric coastlines since the Late Palaeolithic (Figure 6; Bell, unpublished; Tetlow, 2005).

A marine aggregates license was awarded within Area 1 within the last 12 months. Recent survey data exists for this but is still commercially sensitive. Older data exists for the other two licensed areas but is often still pertinent to applications for licence extension. Access to these depends on the agreement of the aggregates producers involved. In the case of the Nash Bank licence, this would involve the agreement of all three operators. The same is true of the environmental and engineering surveys undertaken prior to approval of the Scarweather Sands wind farm.

The University of Wales in Bangor (UWB) part owns a survey vessel capable of being used for 2D seismic surveying, sonar mapping and sediment sampling. They have been involved in surveying Carmarthen Bay in the northwestern part of Area 1. They hold digital data for all surveys conducted since the mid to late nineties. However, staff resources limit responses to requests for data. The NMR of England contains a single record of a Mesolithic artefact having been recovered from within Area 1 (Figure 2.4d). Dawkins and Winwood reported a number of

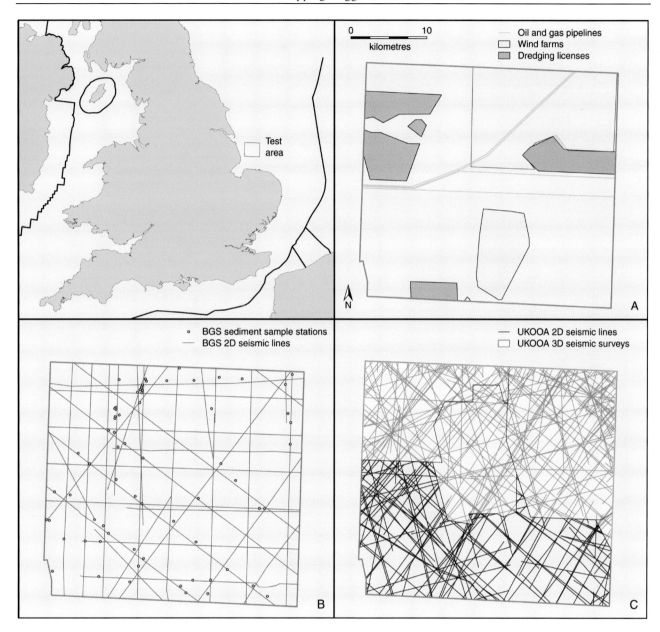

Figure 2.6 Area 3 a) Industrial threats to archaeology b) Distribution of BGS sediment and 2D surveys c) UKOOA 2 and 3D surveys.

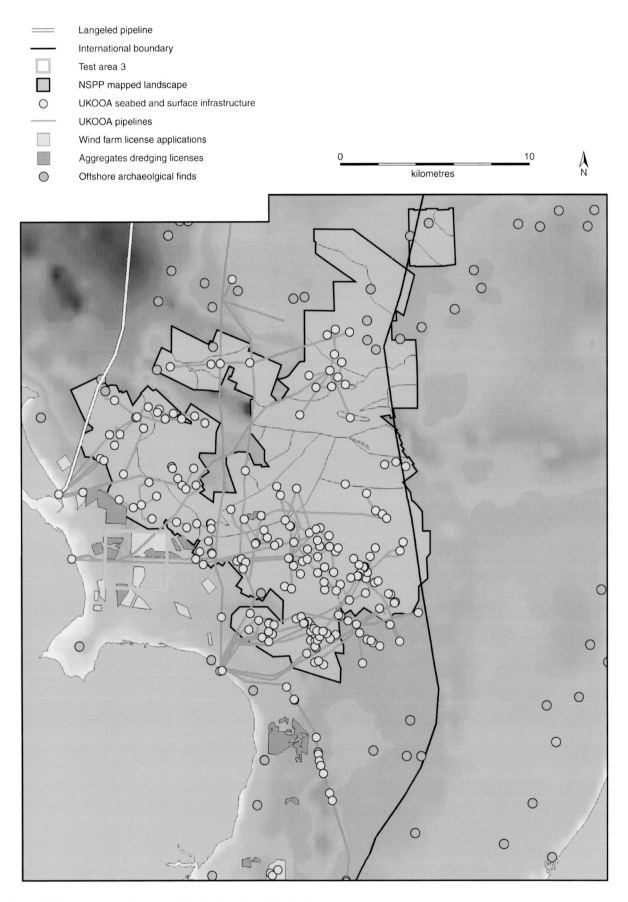

Figure 2.7 Intense use of space within the Southern North Sea

2.6.2 Area 2: Portland

Area 2 has not proven attractive to the petroleum or marine aggregates industries. In 2001 UNESCO inscribed the coastline alongside Area 2 as part of a World Heritage Site – the British Mesozoic Coast. Area 2 is traversed only by 2D geophysics survey lines acquired by the BGS and the petroleum industry. It also contains a few BGS shallow sediment core/borehole sample stations and a single exploratory well bore. Figure 2.5 summarises these survey data.

The petroleum industry 2D seismic survey data within Area 2 belongs to many companies and was shot between the early 1970s and the early 1990s. Development did not follow, although the Maersk Oil North Sea UK Ltd sank well 97/12-1 in late 1995, following extensive surveying including side scan sonar high resolution 2D sub-bottom seismic profiling and vibrocoring, carried out by Kerr McGee Oil (UK).

Three coastal sites are where prehistoric peat and wildlife have been recorded adjacent to Area 2. However, the English NMR and Dorset HER contain no record of Mesolithic finds recovered offshore.

2.6.3 Area 3: The Spurn

Area 3 is situated north west of the Wash and is of particular importance in that it impinges on the NSPP study area. This has been an important area for oil and gas exploration, fishing and shipping and is traversed by gas and chemical pipelines. Recently, it has become a focal point for marine aggregates production and offshore wind farm development. Area 3 occupies part of the "white band" between the current extent of 3D seismic data and the shore (Terrell et al., 2005). However, the area has been intensively investigated by other forms of survey primarily aimed at characterising the shallow seabed sediments.

Figure 2.6b and c illustrate the distribution of BGS and UKOOA survey data within Area 3. A request, dictated by data availability, was made for a strategic selection of seismic lines and shallow sediment core records from the BGS. The BGS was able to supply these as a selection of image files of scanned paper seismic lines and geological reports. Several items from the original enquiry were unavailable. Geological and survey reports for the gas pipeline featured in Figure 2.6a were available from the current owner.

The petroleum industry 2D seismic surveys belong to many different companies. They were shot between the mid 1960s and the early 1990s. Continuing industrial regeneration of this area lends these data considerable ongoing commercial value and, consequently, some may be less easy to obtain.

Half of Area 3 has been investigated by petroleum industry 3D seismic surveys (Figure 2.6c and Figure 2.7). These are part of a group of four surveys fringing the PGS MegaSurvey and belong to WesternGeco. Three are 'unreleased speculative' survey data that would be expensive to obtain. The fourth is a 'released proprietary' survey originally owned by BP Exploration. This should be relatively straightforward to obtain for academic research.

Area 3 contains four zones licensed for shallow marine aggregates dredging, each licensed to a different marine aggregates producer. All underwent the same licence application process despite the fact that one zone, licence 440, lies beyond the 12-mile territorial limit. Technical environmental statements for these zones are not available in electronic form. Any other data that has been retained would prove useful in future licence renewal applications and are likely to be subject to the constraints of commercial sensitivity.

The Greater Wash area has become a major focus for the second generation of offshore wind farm projects. One of two "super" farms currently under consideration by the DTI and DEFRA, Triton Knoll, falls within Area 3. This and the other smaller project, Race Bank, are currently at a late stage of the DTI consents process. Survey data is likely to remain commercially sensitive until 2010.

As area 3 is adjacent to the NSPP study area it is useful to consider a larger area in respect of marine infrastructure. This information is provided in Figure 2.7. Study of the data reveals the intensity of industrial use of the area. Query of the NMR of England, the HERs of Lincolnshire and Norfolk and the literature revealed there have not been any Mesolithic artefacts recovered and catalogued within Area 3. However, a wider, regional view of these combined data suggests that the SNS is a relative hotbed for offshore finds by UK standards. However, Figure 2.7 also suggests a general absence of finds within the region where there is a high concentration of petroleum industry infrastructure. This may well be the result of exclusion zones around rigs and wellheads that prohibit deep fishing and dredging: activities often responsible for disrupting prehistoric archaeological deposits (Wenban-Smith, 2002). Such deposits may therefore have actually been preserved by the presence of this infrastructure. On the other hand, the absence of a unified database of finds supported by sediment or geomorphological data, as provided by the NSPP, is highlighted when one considers that the route of the new Langeled gas pipeline from Norway passed across part of the project study area and, possibly, through a site where moorlog (fossilised peat) was recorded (Kooijmans, 1971, Figure 2.6).

2.7 Conclusions

The requirement for a greater understanding between offshore developers and marine conservationists is becoming more acute as the UK makes greater use of its marine resources in ever more diverse ways. Central to the effective management of the seabed is collaboration between developers and researchers that aims to mitigate our impact on the environment. The archaeological resource contained within the Holocene sediments of the seabed remains poorly understood. This is primarily due to the practical limitations of carrying out archaeological investigations. However, the NSPP has revealed dramatic insights into the nature of the emergent prehistoric landscape around our present-day coast and these insights were gained using existing exploration survey data. The latter point is an important one and is emphasised by the extent and variety of available information revealed as part of this data audit. It should now be clear that much substantive information on the nature of the marine resource could be recovered through the use of the existing data. In some cases the strategic collection of new data may be required, where gaps have been identified or on those occasions where further information is required for specific archaeological purposes.

The final point to be made is central to the use of existing resources. It is acknowledged that geographic distribution of acquired survey data has been led by the location of natural resources and location-specific legislation. However, the physical location of resources and related datasets is equally disparate. Many survey datasets are distributed amongst the archives of the BGS and individual offshore operators. Some data are held within academia but these are often bound by the strictures of specific legal agreements linked to a specific activity. Recent government policy is opening up the flow of data between industry, academia and the public. However, it remains for researchers to prove the usefulness of their work for industry and regulators and to assist in defining the threat that industrial activities pose to the marine environment. In so doing, researchers can help industry meet the demands of an increasingly rigorous regulatory framework.

3 3D Seismic Reflection Data, Associated Technologies and the Development of the Project Methodology.

Kenneth Thomson and Vincent Gaffney

3.1 Introduction

A variety of methods and datasets were potentially available for use in the project. However, as the choice of methods, and hence data types, controls the volume and quality of the results an optimal approach needed to be developed. A crucial consideration was the need to minimise the time involved in the analysis, and hence the expense of the project, whilst at the same time maximising the spatial coverage and detail. Consequently, the available technologies, the costs involved in acquiring new data and the possibilities for using existing data needed to be evaluated when planning the project.

The Southern North Sea (SNS) contains an extensive collection of data including seabed samples, shallow core, bathymetry data and seismic reflection profiles collected for the investigation of near seabed features or deeper hydrocarbon exploration. This suggested that the project could potentially achieve its aims by exploiting existing data with future acquisition of bespoke datasets being considered on the basis of the project's results. The existing datasets from the SNS were acquired for a variety of purposes and consequently have differing strengths and weaknesses. Seabed samples and shallow coring can provide chronological, sedimentological and environmental data. However, such data provides a poor spatial framework.

Although high resolution bathymetry data can provide excellent images of the seabed topography, and hence detailed images of early Holocene features that have a bathymetric expression, many of the important geomorphological features are, at least, partially buried in the SNS. Consequently, there was a need for regionally extensive datasets that have the capability to image below the seabed. The only existing data within the SNS that could meet these requirements were seismic reflection datasets. Use of such surveys could permit the generation of regional maps for buried Early Holocene landscape features. These datasets would then provide the framework into which data from shallow boreholes, seabed samples and bathymetry could be integrated.

Marine seismic acquisition is undertaken for a variety of purposes, with varying data densities, coverage, depths of penetration and resolution. Consequently, there was a choice between differing seismic reflection data types, each being acquired for specific purposes that may not have been compatible with the project requirements. This paper will discuss the critical parameters for differing seismic reflection data types and how these considerations influenced the project methodology.

3.2 Seismic reflection method and resolution

Seismic reflection surveying involves the transmission of acoustic energy into the subsurface and recording the energy reflected from acoustic impendence contrasts. The reflections produced at acoustic impendence contrasts are predominantly the product of changes in lithology with the impendence contrast, or reflection coefficient (the ratio of amplitude of the reflected wave to the incident wave, or how much energy is reflected.), given by the equation:

$$R = (\rho_2 V_2 - \rho_1 V_1)/(\rho_2 V_2 + \rho_1 V_1)$$

Where: R = reflection coefficient
ρ_1 = density of medium 1
ρ_2 = density of medium 2
V_1 = velocity of medium 1
V_2 = velocity of medium 2

With appropriate processing this allows the production of pseudo-depth sections of the subsurface structure with the vertical axis being two-way travel time to the reflector.

Although the basics of this technique are common, the details vary for a range of applications including the investigation of deep crustal structure (Klemperer and Hobbs, 1991), hydrocarbon exploration (Bally, 1987) and near seabed sediment structure (e.g. Salomonsen and Jensen, 1994; Velegrakis and Dix, 1999; Praeg, 2003 and Bulat, 2005). These diverse applications dictate different acquisition parameters that in turn determine the resolution and depth of penetration of the survey as well as the costs involved in acquiring the data. Consequently, the relative merits of a range of available seismic reflection data types needs to be assessed when considering the investigation of submerged, and partially buried Holocene features.

Standard marine acquisition involves towing an energy source and a cable (streamer) containing pressure sensitive receivers to record the reflections from the underlying strata (Figure 3.1). In single fold data, only one reflection is received from any point in the subsurface. However, many seismic profiles are multi-fold. In this case several shot-receiver pairs are of the correct geometry to collect acoustic energy reflected from the same point. These reflections can then be summed in order to increase the signal-to-noise ratio of the seismic profile.

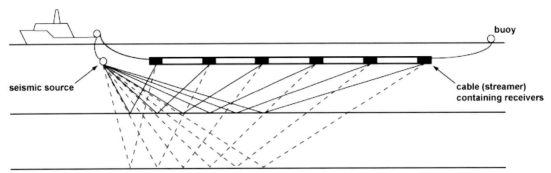

Figure 3.1 Typical marine seismic reflection acquisition. The vessel travels through the water and regularly fires the seismic source. The sound wave travels through the water column (and underlying sediments) and is partially reflected at acoustic impendence contrasts. Receivers within the long towed cable astern of the vessel detect the reflected wave, which is then transmitted to the vessel and recorded.

The characteristics of the seismic energy source are intimately linked to the required resolution and depth of penetration. The vertical resolution of seismic reflection data requires a minimum separation between two interfaces that will give rise to two separate reflections. At separations of less than ¼λ the reflections from the two acoustic impedance contrasts constructively interfere with the maximum amplitude occurring at ¼λ, known as the tuning thickness (Figure 3.2). However, it is not until ½λ that the two reflections are separable (Figure 3.2). Consequently, the vertical resolution of seismic reflection data can be defined as either the minimum resolvable (¼λ; Figure 3.3) or the minimum separable (½λ; Figure 3.3).

As the vertical resolution is dependent upon the wavelength it is therefore dependent on the velocity of the medium and the frequency of the seismic source/reflected wave. Ideally, a high frequency source (>100Hz) would be used in all circumstances. However, as the geology progressively dampens high frequency seismic signals with depth, the seismic source needs to be chosen with consideration to the required depth of penetration. The dampening effect of the top few hundred metres of overburden is relatively small and consequently seismic sources with frequencies in excess of 100Hz can be employed. In contrast, 2D and 3D seismic data acquired for hydrocarbon exploration need to image to depths of several kilometres and consequently employ sources with frequencies of less than 100Hz. This, combined with increasing velocity with depth, results in a significantly higher vertical resolution for 2D seismic data specifically acquired for the investigation of shallow geology (<1km) compared to standard 2D or 3D seismic data required for hydrocarbon exploration. This is demonstrated in Figure 3.4 where a high frequency 2D seismic line specifically designed to image Holocene and Pleistocene features is compared with a line from exactly the same position extracted from a 3D seismic dataset acquired for hydrocarbon exploration. The high-resolution line (Figure 3.4a) shows a channel and its complex infill pattern. In contrast, the low-resolution 3D seismic line (Figure 3.4b) is unable to image the channel.

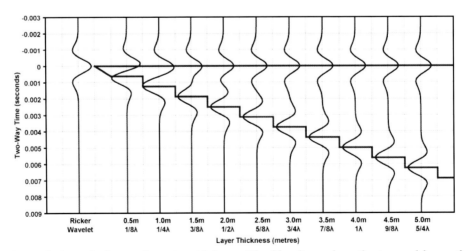

Figure 3.2 Seismic resolution of a layer of varying thickness. The blue lines show the top and base of a layer. Using a 400Hz Ricker wavelet, and assuming a sediment velocity of $1600ms^{-1}$, the reflections from the top and base can be shown to constructively interfere at thicknesses less than ¼λ (1 metre) with the maximum amplitude at ¼λ. Furthermore, the top and the base of the layer do not perfectly align with peak and troughs. At thicknesses of ½λ (2 metres) and greater the top and the base of the layer are separable with peak and troughs aligning perfectly with the position of the top and base respectively.

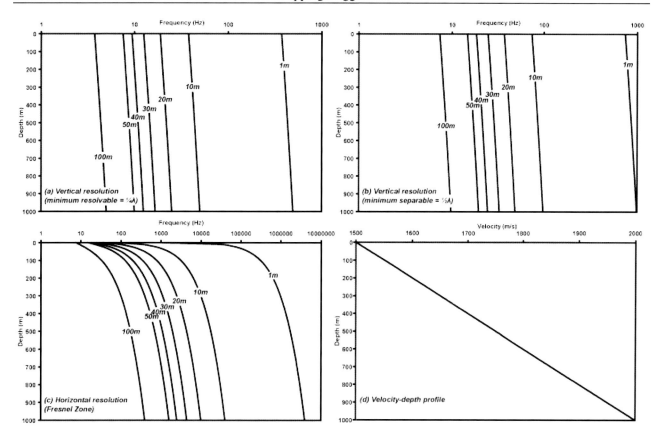

Figure 3.3 Plots of seismic resolution as a function of burial depth and frequency. (a) The minimum resolvable vertical resolution. (b) The minimum separable vertical resolution. (c) The horizontal resolution of unmigrated seismic data. (d) The assumed velocity-depth structure.

The lateral resolution of seismic reflection data is dependent on the fact that seismic energy travels through the subsurface and encounters the reflecting surfaces over discrete areas. The energy travels as wave fronts and the region on the reflector where the seismic energy is reflected constructively is known as the Fresnel Zone (Sherrif, 1977). Lateral resolution is determined by the radius of the Fresnel Zone, which itself depends on the wavelength of the acoustic pulse and the depth of the reflector. Thus, in non-migrated seismic data, lateral resolution is dependent on the frequency of the seismic source, the interval velocity and on the travel time to the reflector. As with the vertical resolution, this implies that high frequency seismic reflection data will provide a significantly higher lateral resolution compared to lower frequency data collected for hydrocarbon exploration (Figure 3.3).

However, the procedure of migrating seismic data, which ensures reflected energy is correctly positioned within the subsurface, considerably enhances resolution. Consequently, for migrated data, lateral resolution depends on trace spacing, the length of the migration operator, time/depth of the reflector and the bandwidth of the data. If completely successful then the lateral resolution of the high frequency seismic section shown in Figure 3.4a would be approximately 12m compared to 50m for the low-resolution 3D seismic line (Figure 3.4b).

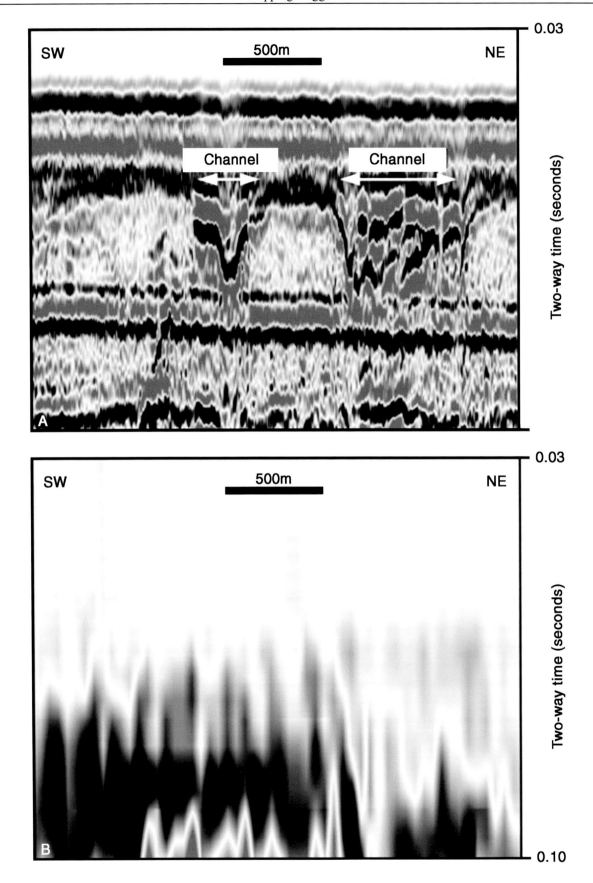

Figure 3.4 A comparison between (a) high frequency 2D seismic reflection line and (b) low frequency 3D seismic line from the same location. Note that higher frequencies yield greater vertical detail.

3.3 2D versus 3D seismic acquisition and interpretation

Traditional seismic reflection data is generally referred to as 2D as it is acquired as a series of discrete vertical profiles using a single streamer towed behind the vessel. This acquisition pattern results in the collection of several profiles with the spacing between profiles being several orders of magnitude greater than the trace spacing (i.e. the horizontal sampling interval along the profile).

This method of acquisition has two main disadvantages. Firstly, the reflected seismic energy is assumed to have originated from a point directly beneath the profile even though it could have originated from a point laterally offset from the profile. This aliasing means that the location of a feature cannot be accurately constrained, as the spacing between lines is too wide to correct this error. Secondly, the spacing between lines is sufficiently wide that it can be difficult to map the position of a morphological feature across the region of interest. For example, Figure 3.5 demonstrates how wide line spacing can lead to several equally valid interpretations.

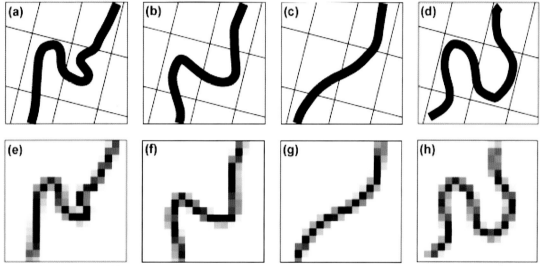

Figure 3.5 (a-d) Four possible interpretations of a channel morphology based on a coarse 2D seismic grid. Each interpretation is equally valid. (e-h) Schematic illustrations of how each of the interpretations shown in a-d would appear on a timeslice from a laterally continuous, binned 3D seismic volume. This demonstrates that 3D seismic data has the potential to distinguish between the possible alternatives.

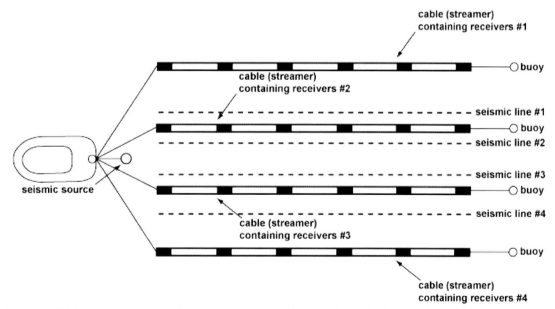

Figure 3.6 Typical 3D marine seismic reflection acquisition. The vessel travels through the water and regularly fires the seismic source. The sound wave travels through the water column (and underlying sediments) and is partially reflected at acoustic impendence contrasts. Receivers within the towed cables astern of the vessel detect the reflected wave, which is then transmitted to the vessel and recorded. In contrast to Figure 3.1, 3D acquisition involved multiple towed cables with each cable being capable of producing a seismic profile.

In contrast, 3D reflection seismic data involves the towing of multiple streamers (Figure 3.6), which allows the rapid collection of multiple closely spaced lines. The survey configuration provides significant advantages as it generally involves multifold collection, which is containing several reflections from the same subsurface location, and allows reflected energy to be correctly positioned in space. Thus it eliminates the potential positioning errors associated with 2D seismic data. This is achieved through the conversion into a binned dataset with, in the case of data acquired for hydrocarbon exploration, a bin spacing of 12.5m x 12.5m x 4 milliseconds, or multiples thereof.

Each bin is then populated by those reflections that originated from within the bin space. Once binned the data format provides additional benefits. Firstly a laterally continuous binned data volume means that a geomorphological feature can be mapped from bin to bin, removing the potential errors involved in the interpretation of 2D data (Figure 3.5 and Figure 3.7). However, 3D seismic is also more versatile than 2D data as it can be interrogated in a number of ways. Instead of relying on vertical profiles, the volume can be sliced in any direction. Of particular importance to the investigation of relatively shallow, and flat, Holocene features is the ability to produce a horizontal slice (timeslice) through the data as this can, in many cases, be interpreted as a geological map showing a range of sedimentary features and facies patterns (Figure 3.8; Fitch et al., 2005).

The interpretation of 3D seismic data has improved significantly in recent years due to the development of a range of new techniques originally designed to improve geological interpretation associated with hydrocarbon exploration and production. Once a stratigraphic marker of interest has been identified, it can be mapped across the 3D seismic volume to produce a horizon. This can then be examined using a variety of attributes (e.g. depth, seismic amplitude, dip, azimuth) with each attribute having the potential to reveal different characteristics of the feature of interest (Posamentier, 2005). However, simply applying artificial illumination from a number of directions can prove highly effective at identifying subtle geomorphological features.

Seismic attributes can also play a crucial role in the interpretation of 3D seismic data. Extracting RMS amplitude (root-mean-square) over a two-way time interval, defined by two mapped horizons or two timeslices, is commonly employed to differentiate zones of different seismic amplitude within a seismic volume (e.g. Den Hartog Jager et al., 1993). As seismic amplitude is a function of density and/or velocity contrasts it is often closely related to the depositional facies (Figure 3.8). Further enhancement of the seismic slices, both vertical and horizontal as well as mapped horizons, can be achieved through the generation of a coherence (or semblance) seismic volume. The coherence cube (Bahorich and Farmer, 1995) calculates localised waveform similarity in both inline and crossline directions and estimates of three-dimensional seismic coherence are obtained. Small regions within the seismic volume containing stratigraphic anomalies such as channels have a different seismic character compared to the corresponding regions of neighbouring traces. This results in a sharp discontinuity in local trace-to-trace coherence and allows the rapid identification of stratigraphic features (Figure 3.8; Bahorich and Farmer, 1995).

Another advance in 3D seismic interpretation has been the development of opacity rendering techniques (Kidd 1999). The technique converts conventional 3D seismic data into a voxel volume. Each voxel contains information from the original portion of the 3D seismic volume that it occupies together with an additional user-defined variable that controls its opacity. The opacity of individual voxels can then be varied as a function of their seismic amplitude (or any other seismic attribute), allowing the user to examine only those voxels that fall within the particular amplitude (or attribute) range of interest. By using appropriate opacity filters it is possible to image the depositional systems such as buried fluvial channels. This exploits seismic characteristics, which are in part lithologically dependent, and different from the surrounding materials, thus permitting the surrounding rock to be made transparent whilst preserving all but the smallest channels as opaque features (Figure 3.8; Fitch et al., 2005).

3.4 Interpretation strategy for the Southern North Sea

The above discussion demonstrates that the ideal dataset for the investigation of submerged Holocene/Mesolithic landscapes within the region would be high resolution (>100Hz) 3D seismic data with appropriate borehole control. Such a dataset would provide high (metre or less) vertical and lateral resolution and a laterally continuous coverage, thus removing the need to interpret the location of a feature between data points. It would also support the application of the latest advances in seismic attribute analysis and visualisation to aid interpretation Unfortunately, high-resolution 3D seismic surveying equipment, such as the 3D CHIRP system developed by the National Oceanography Centre, Southampton (Gutowski, 2005), is a recent development and consequently such data is extremely rare. Furthermore, such systems currently utilise small vessels that are not suitable for deployment beyond the immediate coastal waters. However, it is the high resolution of the system that is the most significant handicap. High-resolution seismic acquisition involves much slower surveying rates and thus higher costs. For example, the 3D CHIRP system described by Gutowski (2005) is capable of surveying approximately $0.02km^2$ per day. This contrasts with the lower resolution 3D seismic data acquired for hydrocarbon exploration which, although it involves large, expensive and custom-built vessels, can survey $40km^2$ a day at a cost of c. \$5000 per square kilometre (Bacon et al., 2003).

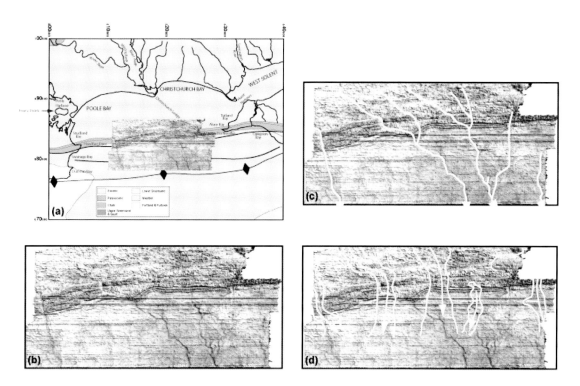

Figure 3.7 (a) Geological map of Poole and Christchurch bays with the artificially illuminated plan view of the seabed reflector. (b) Artificially illuminated plan view of the seabed reflector. (c) Artificially illuminated plan view of the seabed reflector with palaeo-channels as mapped using 3D seismic. (d) Artificially illuminated plan view of the seabed reflector with palaeo-channels as mapped using 2D seismic data (Velegrakis et al. 1999). Note the poor correlation between the 2D interpretation and the actual channel locations.

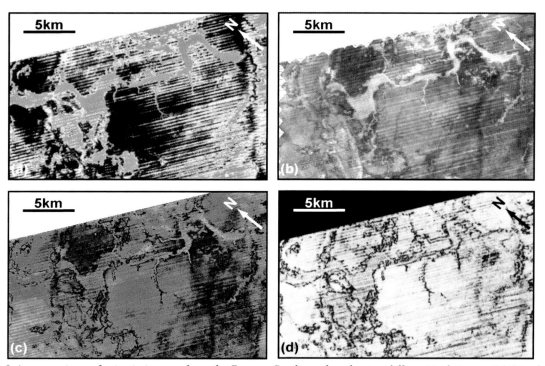

Figure 3.8 A comparison of seismic images from the Dogger Bank produced using different techniques. (a) Simple seismic amplitude timeslice revealing some Holocene depositional features (channels). (b) RMS amplitude extraction for the top 200ms of the data. (c) Opacity rendering of the same seismic volume as (b). Note that both the RMS extraction and the manipulation of the opacity allow some improvement in channel definition. (d) Seismic coherence (semblance) timeslice at the same level as (a). Note that coherence provides some additional detail of the channels.

Although high-resolution 3D seismic data is expensive to acquire and not readily available it is interesting to consider the advantages of using existing 3D seismic data acquired for hydrocarbon exploration. The frequency spectra for the top 200ms (the likely depth/two-way time range of interest) of the dataset used in this study has 98.7% of the frequency content in the 3-72Hz range with a mean frequency of 14.7Hz. Consequently, a mean frequency of 14.7Hz provides a vertical resolution of 27m although the higher frequency components suggest that a vertical resolution of 10m or less may be possible. The limit of horizontal resolution for unmigrated seismic data, the Fresnel Zone (Sherrif, 1977), for the mean frequency of 14.7Hz would provide a Freznel Zone width of 66m. However, this can be considered an extremely conservative estimate as the higher frequency components suggest lateral resolutions of around 30m may be possible (Emery and Myers, 1996). In addition, as migration of the seismic data considerably enhances lateral resolution with the limit being dependent upon trace spacing, length of the migration operator and the bandwidth of the data. Consequently, the lateral resolution of the top 200ms of the migrated dataset used in this study may actually approach the line spacing of 50m.

Although these parameters may suggest that the commercial 3D seismic datasets acquired for hydrocarbon exploration are not suitable for the exploration of Holocene/Mesolithic landscapes in the Southern North Sea the reverse is actually true. For example, Figure 3.7 is an artificially illuminated map of the seabed reflector from Poole and Christchurch bays in the English Channel and mapped using commercial 3D seismic data from over the Wytch Farm oilfield. The map shows a number of north-south trending channels that do not possess a current bathymetric expression. Instead, Velegrakis et al. (1999) demonstrated, using 2D seismic data, that these channels are completely infilled and are sub-seabed features. Consequently, Figure 3.7 demonstrates that the vertical resolution of approximately 10m results in a mean response from both the seabed and the acoustic impedance contrasts from several metres below it (c.f. Bulat, 2005). This implies that mapping the seabed, or any other near seabed reflector, can provide maps containing information from several metres of Holocene strata. Furthermore, a timeslice can also be considered to provide information from a stratigraphic interval several metres thick. Given a bin spacing of 50m, and an areal coverage of >20,000km^3, this suggests that timeslicing and mapping regionally significant reflectors has the potential to provide an extensive reconnaissance tool for the investigation of submerged, Holocene landscapes (Figure 3.8).

The above considerations therefore suggest that an alternative interpretation strategy could be employed for the investigation of Holocene/Mesolithic landscapes beneath the Southern North Sea. This approach was completely dependent on the donation of >23,000km^3 of commercial 3D seismic data by PGS Reservoir. This would provide an opportunity to rapidly develop a regional framework into which other, higher resolution datasets, could subsequently be integrated. The approach was:

1. To map regionally significant reflectors using the regional 3D seismic dataset.
2. To interpret these surfaces using artificial illumination and horizon attributes such as amplitude and dip to identify morphological features and the developmental chronology.
3. To generate seismic attributes for the regional 3D seismic dataset.
4. To sequentially timeslice these attribute volumes (e.g. amplitude, coherence, RMS amplitude) and to employ opacity rendering techniques to identify morphological features and the developmental chronology.
5. Integrate the above to develop a first order geomorphic model.
6. Use existing high resolution 2D seismic data and shallow borehole data to refine the geomorphic model, resolve interpretational and chronological ambiguities, and to provide palaeoenvironmental data.

This strategy provided several advantages. Firstly, it optimised the use of existing data thus reducing the cost of the project. More importantly, the speed of the project was significantly increased as the 3D seismic permitted the rapid development of a regional model and the identification of key localities for detailed work. These could be identified on the basis of the need for clarification or a recognition of their environmental importance.

3.5 Conclusions

The key requirement for understanding the Mesolithic archaeological potential of the Southern North Sea is the development of a detailed regional landscape model. Given the vast area under consideration the traditional archaeological approach would suggest a requirement for the acquisition of a large bespoke geophysical survey of the area, with appropriate stratigraphic, sedimentological and environmental controls from cores. The expense, and logistical complexity, involved in such a survey would be prohibitive. Consequently, there existed a need to develop a regionally extensive and detailed landscape model utilising existing data. Fundamental to developing such a model would be access to regionally extensive seismic reflection datasets. However, given the range of uses for which data are acquired, the acquisition parameters vary considerably and hence have varying degrees of applicability to the study of Early Holocene landscapes. Traditional 2D high frequency, high resolution seismic reflection data have been the preferred dataset for the investigation of Holocene geology and archaeology. Unfortunately, such datasets are limited in their use as regional mapping tools as they are prone to spatial aliasing errors and require extrapolation of geomorphic features between relatively widely spaced data points. Conversely, 3D seismic data acquired for hydrocarbon exploration have significantly lower resolution but provide a relatively complete spatial coverage. This diminishes the aliasing issues and also allows a range of techniques such as timeslicing, attribute analysis and seismic visualisation to be applied. These advantages permit the rapid development of a regional geomorphic model into which higher resolution seismic reflection data, bathymetry, seabed samples and shallow core can be integrated.

4 Merging Technologies: Integration and Visualisation of Spatial Data

Simon Fitch, Vincent Gaffney and Kenneth Thomson

4.1 Introduction

The primary goal of the North Sea Palaeolandscapes Project (NSPP) was to explore the Late Pleistocene and Holocene landscapes of the Southern North Sea through the use of c. 23,000 km² of contiguous 3D seismic data provided by PGS UK (www.PGS.com). Processing this large volume of data required considerable investment in both hardware and software. In line with most archaeological projects utilising large amounts of digital, remote sensed data, there was a requirement to utilise specialist softwares for the processing of primary seismic datasets and standard geographic information systems to manage, manipulate and display supporting data and interpretative layers (Gaffney and Gater, 2003 Chapter 5; Chapman 2006). What is, perhaps, less usual within archaeology has been the requirement to provide access to high bandwidth networks and storage to cope with the large volumetric datasets used by the project and to provide access to high-end stereo-projection systems to visualise and to quality check the data as it was processed and interpreted. Given the nature of the project and its significant technical demands some description of the technical context of the project is required. Technical management of the project ran through the HP Visual and Spatial Technology Centre (HP VISTA). HP VISTA is a division of Birmingham Archaeology and is situated within the Institute of Archaeology and Antiquity at the University of Birmingham. The laboratory was developed, exceptionally, for the visualisation of large archaeological data sets and the facilities provided are, currently, unique within British archaeology. A schematic of the Centre's specialist infrastructure, as utilised by the NSPP, is provided in Figure 4.1.

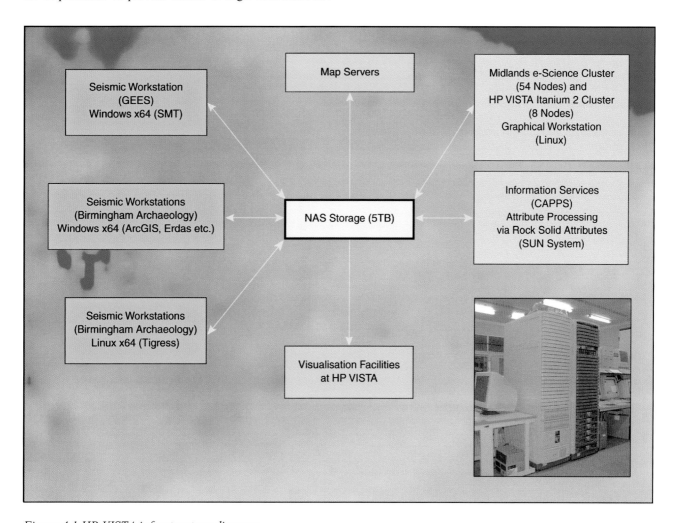

Figure 4.1 HP VISTA infrastructure diagram

4.2 Infrastructure

The primary engines for the majority of analysis were three HP xw8200 workstations with 64bit dual processors, 16Gb of RAM, 2Tb of local storage and high-end graphics cards. Whilst the majority of primary processing was carried out on these machines, specialist softwares used to carry out attribute analysis on the seismic data, for example, were resident on machines elsewhere on the campus. Project data was held centrally on a secure 5Tb NAS storage system resident within the University Information Services' machine room. The need to move data between these machines and to the central storage was supported by gigabit-enabled cabling between most project computers, and assisted by the availability of a dedicated fibre optic link between the NAS and the main display centre in the HP VISTA centre. The extent and complexity of the volumetric data, and the utilisation of solid modelling and opacity rendering techniques, required that the data should be viewed stereoscopically and also that displays were large enough to provide the opportunity of group viewing. This was provided within the HP VISTA centre via an HP SV7 scaleable visualisation system. This is a multi-pipe system providing high-end, high quality visualisation. A particular benefit of the system is the ability to set the level of performance, image quality and resolution independently, so a large model can be rapidly manipulated at a lower quality/higher performance setting and then a particular region of interest examined at a high quality/lower performance setting, without having to interrupt the visualisation. The SV7 supported projection across a 4.27m by 1.8m dual channel, rear projection Fakespace Power-Wall. This provided a geometrically accurate stereoscopic display using active stereo glasses (Figure 4.2). This combination of analytical power and sophistication of display, rare within archaeology, was demanded by the nature of the datasets under investigation and is comparable with specialist facilities dealing directly with petroleum geology and remote sensing data sets. To put this in context, a paper published 10 years ago and also concerned with the issues of integrating GIS and other spatial technologies within archaeological projects at Birmingham, records that the entire archaeological computing group was then serviced by a single server with 64 megabytes of RAM and 4 gigabytes of mass storage (Buteux et al. 1997, Gaffney and Gaffney 2000a). The NSPP's computing requirements, and its demand for exponentially increased storage capacity, presumably indicates the scale of change that can be anticipated in future, comparable digital landscape projects.

Figure 4.2 Inspecting data at the Visual and Spatial Technology Centre (HP VISTA)

4.3 Software integration

Having acquired the infrastructure needed to implement the project it is appropriate to consider the requirement for software integration that arose during the course of the study. The increasing application of geographic information systems to manage, manipulate and display spatially referenced data sets within archaeology has been a major trend over the past decade (Chapman 2006, Connolly and Lake 2006). Not surprisingly, the majority of the NSPP's interpretative and supporting spatial data has been held within an ARCGIS database made available to all team members. A similar trend is discernable within other comparable groups concerned with hydrocarbon exploration. The ability of GIS technology to handle a variety of spatial data, in conjunction with its analytical capacity, make GIS an invaluable tool for petroleum exploration (Gaddy 2003). Further, the ability of a GIS to visualise and manage data throughout a project's life has also proved invaluable to the petroleum industry (c.f. Lawley and Booth, 2004).

Despite this promising situation, the size of the project database and the incorporation of volumetric data sources proved problematic during the course of the research. It is notable that the application of GIS systems within the petroleum industry, in contrast to archaeology, has run in parallel with the development of 3D modelling systems and remote sensing packages that link interpretive and geophysical data (Gaddy 2003, 1). The requirements of the petroleum industry have produced highly flexible interpretation packages and softwares, such as Tigress and SMT Kingdom, developed to image vast, complex geophysical datasets and to facilitate the mapping, management and planning of petroleum data within an easily visualised environment. Industry requirements are therefore very similar to those of the NSPP (see Thomson and Gaffney, this volume), and the utilisation of these developed and reliable packages was highly desirable in the context of the analysis carried out as part of this project.

Within marine archaeology, the use of GISs to assist in archaeological management is fundamental (Groom and Oxley 2001, 56). Distributional analysis of marine and associated resources, or analysis of absence of evidence, permits targeted use of resources in curatorial terms and provides a greater insight into the structure of the marine database (Groom and Oxley 2001; Allen and Gardiner 2000, Fitch et al. 2005, 194). The utilisation of geophysical information in this role within a GIS is also appreciated by the archaeological computing community (e.g. Buteux et al. 2000, Gaffney et al. 2000b). This has proved invaluable when monitoring landscapes that contain poorly understood archaeological resources (Chapman et al. 2001). In projects, such as the NSPP, where archaeological survey and interpretation is severely limited by the prevailing physical environment, the ability to combine data sources including geophysics, physical samples and findspots to provide a proxy environment for interpretation is crucial. The use of GIS to explicitly integrate traditional land based geophysical surveys (e.g. magnetometry and resistivity) as well as remotely sensed imagery is also reasonably well established (Gaffney et al. 2000). The planar nature of these traditional land based geophysical surveys facilitates their easy integration into a traditional GIS's map style interface. However, the representation of true three dimensional volume data, including 3D seismics, is a challenge to systems that are, essentially, 2D (Kvamme 2006, Watters 2006).

Whilst the representation of a third dimension is possible within certain GIS viewers, such applications are not ideal for the purposes of representing volumetric geophysical data, as these do not allow for the representation of voxels. Such systems are therefore unable to adequately display a volumetric representation of a cube of seismic 3D data. Within the NSPP there was, therefore, always a requirement to explore the means by which volume data could be integrated with standard GIS map layers in a manner that retained the integrity of the original volumetric data.

4.4 Primary integration procedures

Whilst proprietary tools do exist to display the products of the analysis of seismic data within a GIS, this has generally been achieved through the rectification of a flat image from which interpretation can be undertaken. With respect to seismic surveys, integration of a cube of data can be achieved thought the export of serial planar, timeslices. This facilitates interpretation alongside GIS layers in a traditional fashion (Goodman and Nishimura, 2000). However, for the NSPP it would not be desirable to introduce all possible time slices into the GIS. Indeed, to display the number of seismic attributes used in this project would result in tens of thousands of slices and associated data layers. This would add unnecessary complexity to the database and vastly increase the amount of data to be manipulated and stored. Indeed, management of such a volume of data would probably be beyond the present capacity of most GISs and, probably, incomprehensible to a human operator (Kvamme 2006).

Consequently, in the first instance a selection of slices from the most commonly used data attribute amplitudes were selected to provide an overview of the dataset. This selection, however, does leave open the possibility that significant information may be missed between slices. To compensate for this the NSPP utilised RMS amplitude slices (root mean squared, see Thomson and Gaffney, this volume) to facilitate the display information from the missing volume within a standard 2D image. The resultant slice therefore shows areas of anomalous seismic amplitudes within the selected volume, and can be useful in imaging channels (see Figure 4.3). The resultant output is a planar slice, admirably suited for the integration into a standard GIS system, and satisfies some of the requirements of Kvamme (2006) for displaying this data type. Furthermore, the method of generation means that it is highly accurate in terms of spatial location, and is only limited by the properties of the original seismic survey.

Thus such an approach is compatible with standard processing and display methodologies used in archaeological prospection (see Watters (2006) for a comparable process used to process GPR data).

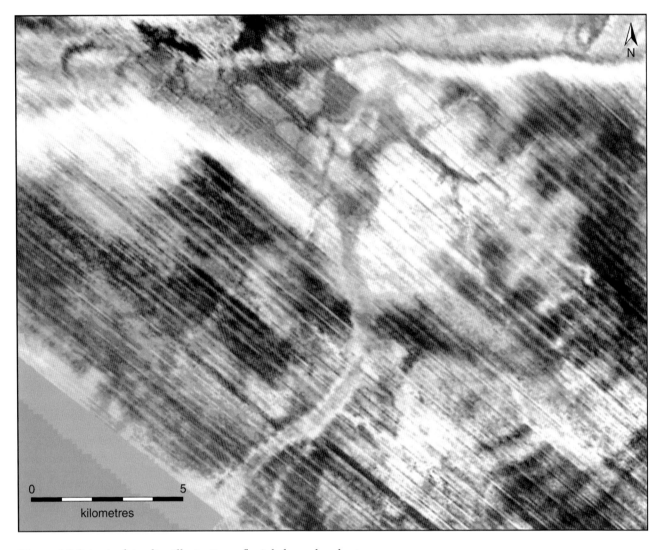

Figure 4.3 Seismic data slice illustrating a fluvial channel and estuary

4.5 Integration of volumetric information through solid modelling

Whilst RMS slicing allows for the display of more of the information contained within the volume of seismic data, its generation as a planar slice results in the loss of integrity of complex structures because the three dimensional component of any anomaly is not adequately represented (Figure 4.4). The potential to lose significant data is therefore very real. From the outset of the project it was decided to follow common practise amongst petroleum geology groups and to use the sophisticated display technologies available at Birmingham to implement analyses permitting solid modelling and full 3D and stereo visualisation to maximise information extraction.

3D surface modelling has been utilised within archaeological geophysics for some time (e.g. Neubauer and Eder-Hinterleitner, 1997). Yet, fundamentally, these still represent only a single layer and are, at best, 2.5 dimensional in nature. They are not, therefore, suitable for the representation of the volumetric 3D seismic data or for the exploration of an internal structure within complex volume features.

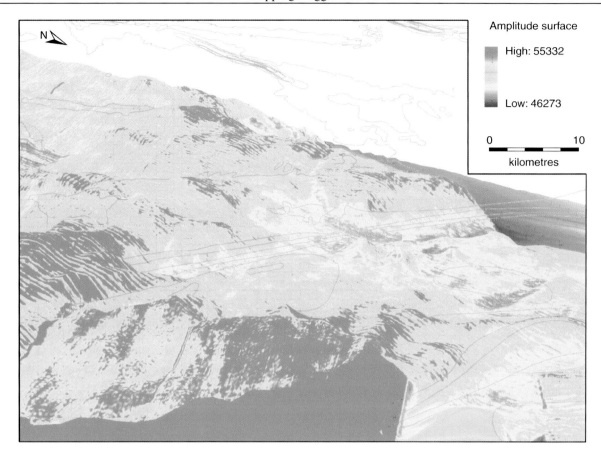

Figure 4.4 3D amplitude surface within a GIS. The green lines are bathymetric contours, and allow for the visual correlation between the anomalies and seabed topography.

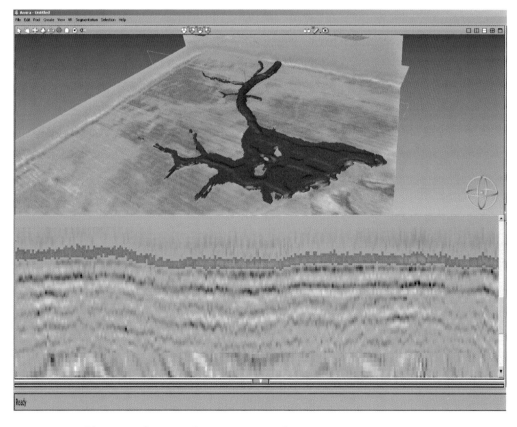

Figure 4.5 Segmentation of features of interest from a seismic volume.

Mercury Software's 'Avizo' visualisation package was utilised within the NSPP to provide a visual insight into the volumetric components of anomalies observed within the 3D seismic data (Figure 4.5). This was supplemented by purchase of the Avizo "Very Large Data Pack" which supports datasets that may be hundreds of gigabytes in size. Representation of anomalies from industry standard SEG-Y data is possible utilising the recent features in Avizo software developed to facilitate geological visualisation services for the oil and gas industry. Following import, Avizo possesses a suite of tools, which make it an effective environment to explore, analyse and display many types of remotely sensed data. The potential to specifically extract information associated with voxels within a series of solid models using this software has recently been demonstrated by Watters using ground penetrating radar data (2006).

To extract the required volumetric models, the original seismic data was directly segmented utilising picking techniques that are commonly employed in the oil and gas sector (Figure 4.5). A series of user determined lists are generated following segmentation, which contain features of interest. Fully automated selection tools can be utilised to define the voxels contained within each feature. Although this can be an automated process noise, at survey boundaries and within the top of the data column prevented such a simple implementation. Consequently, it was necessary to utilise a semi-automatic process of iso-surfacing, utilising user set boundaries and thresholds to constrain the process of automatic voxel selection. With the required information extracted from the seismic dataset, a solid model was built for each segmented feature by generating a three dimensional frame for each feature model (Figure 4.7). Once constructed, it is possible to disassemble a feature through slicing and gain further insights into the three-dimensional structure (see Figure 4.8). Further information can also be gained through volumetric analysis within the solid modelling package. For example the determination of an average voltex value within a volume according to assigned values or attributes may have particular significance. The calculation of channel volume is an obvious output from such a model. However, whilst valuable in their own right such modelling packages are constrained, in analytical terms, as they lack many of the spatial procedures available to the majority of GISs.

Figure 4.6 Wrapping of identified features within seismic data

Figure 4.7 Solid model generated by wrapping

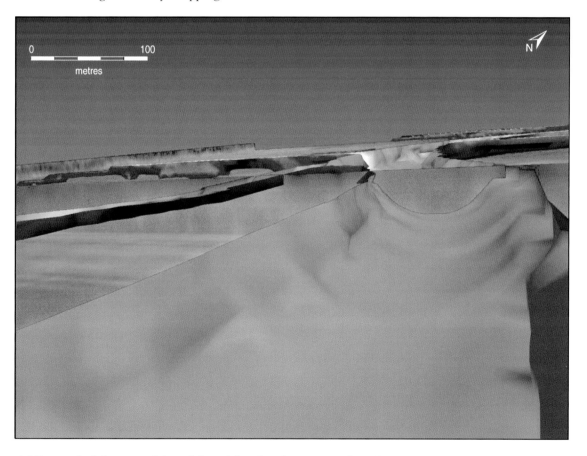

Figure 4.8 Removal of elements of the solid model within the Avizo package from Figure 4.7 permits visualisation of the internal structure of the system

As the volume forming the solid model has real world attributes in all three dimensions it becomes possible to generate CAD models that can be imported into a GIS for display and spatial analysis. This mode of representation is more suited to GIS as it is composed of polygonal elements, which, as they wrap the entire anomaly, can visualise and facilitate the user's appreciation of the actual volume and size of anomalies. The results of part of such a process can be seen in Figure 4.9. However, although export of these models from a fully three-dimensional environment into a GIS permits a representation of the shape and size of the anomalies, it still fails to represent the attributes of the anomaly and the volume of the original survey. Although it is possible to transfer attribute data to GIS data layers, solid modelling remains a superior method to represent 3D anomalies contained within volumetric data and can be used in preference to standard planar slices.

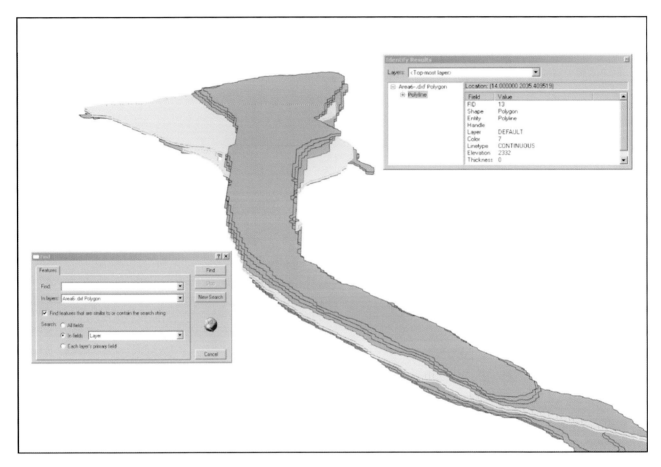

Figure 4.9 Exported solid model within the GIS system. The feature can be seen to be composed of a contour model and a series of shapes representing the geophysical anomaly at the given time slice.

4.6 Merging technologies

Commercial software groups have appreciated the increasing importance of GIS data and the requirement for integration with specialist softwares and data. 3D seismic interpretation packages, including Tigress and SMT's Kingdom, have the ability to create and display datasets derived from, or exported to, a separate GIS. Consequently, GIS layers can be integrated within a fully 3D environment which provides the capacity to display the voxel volume of the dataset and also vector or polygon features derived from that data (Figure 4.10). The ability to display associated GIS information, and the ability to produce GIS interpretation layers directly within the package helps reduce spatial positioning errors, which may occur during traditional planar slice integration. The direct generation of GIS-compatible interpretative layers within specialist packages also improves quality of the interpretation through the availability of advanced attribute and opacity rending techniques not available within a standard GIS (see Thomson and Gaffney, this volume).

Figure 4.10 GIS Layers display within a fully 3 Dimensional environment - A cube of seismic data (displayed as a voxel volume) and a 3D surface & 3D GIS polygons (interpretation layer)

4.7 Conclusions

It is apparent that the use of solid modelling packages to display and analyse seismic volumetric data can assist in integrating complex 3D data within environments that utilise standard GISs for data management. However, it must still be acknowledged that the process of exporting data divorces the solid model from the original volumetric datasets and, potentially, important attribute data or specialist attribute derivatives. Ultimately, this must limit the analytical potential of derived data. However, integration of volume data within a GIS provides enhanced opportunities for spatial analysis plus integration with other, supporting spatial datasets and this cannot be achieved adequately within any proprietary seismic processing package. Until the technologies merge further, and this is a real trend in software development, linking disparate data types through alternative technologies, including solid modelling softwares, offer the best way forward for the integration and visualisation of information derived from volumetric geophysical survey and GIS.

Despite this, it is accepted that the integration of all data sources utilised by archaeological, or related, projects is probably not feasible and, perhaps, may not even be required at this point in time. However, what is incontrovertible is that the complexity of our analyses demands that we are able to transfer the rich spatial data that we utilise between technologies that permit us to visualise them in an appropriate and increasingly sophisticated manner. Indeed, this may ultimately be the most significant point. Our ability to visualise data is increasingly a primary driver and is linked directly into novel interpretative positions. When considering 3D data sets specifically, it is our capacity to facilitate visualisation, through linkage between diverse softwares, which allow us to view data in a variety of new and exciting manners. Within the larger context of available spatial data sources it is this process that will progress our interpretation of novel data sets and, ultimately, our understanding of past landscapes.

5 A Geomorphological Investigation of Submerged Depositional Features within the Outer Silver Pit, Southern North Sea

Kate Briggs, Kenneth Thomson and Vincent Gaffney

5.1 Introduction

A distinct east-west trending bathymetric deep, the Outer Silver Pit (OSP), lies at approximately 54°N 2°E on the bed of the North Sea (Figure 5.1). This deep is the largest of a series of offshore depressions in the Southern North Sea (SNS) that are thought to have formed during Quaternary glaciations, either as the product of subglacial processes (e.g. Valentin, 1957; Robinson, 1968; Balson and Jeffery, 1991; Praeg, 2003) or catastrophic drainage events in an ice marginal environment (Wingfield, 1990). Alternatively, Donovan (1965) postulated that strong tidal currents in the SNS during the early Holocene marine transgression were responsible for eroding such deeps.

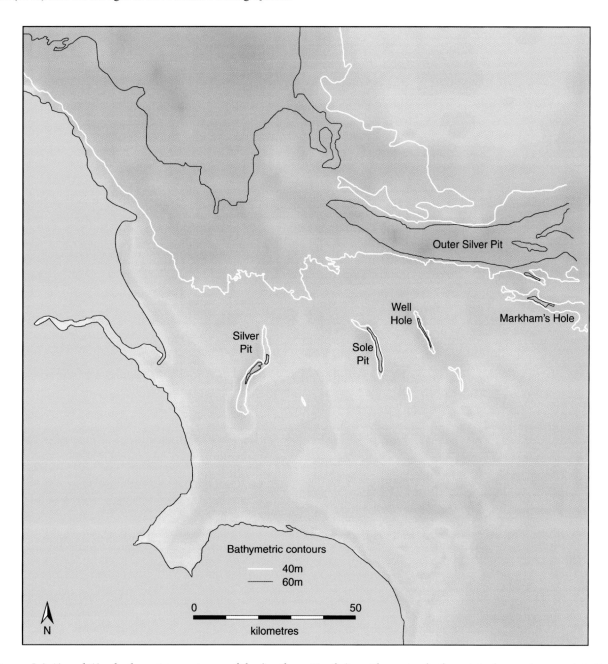

Figure 5.1 40 and 60m bathymetric contours of the Southern North Sea. The major bathymetric depressions are labelled.

Whilst there has been considerable research into the formation of the OSP, the localised geomorphology within the depression has been largely ignored despite its potential to provide further insights into the processes that led to the formation of the depression. The availability of 3D seismic data to this study provided an ideal dataset for the investigation of the morphological and stratigraphical characteristics of the OSP. This chapter will demonstrate that 3D seismic data can be used to identify a number of distinctive geomorphological features within the OSP, the most prominent of these being two elongate ridges that also have bathymetric expression (Figure 5.2). These features will form the focus of this paper. It is the aim of this study to classify the elongate ridge features, and extract any information about the environments and conditions in which they were formed, through close examination of their morphology and locality, and by detailed comparison to modern bedforms with similar morphology.

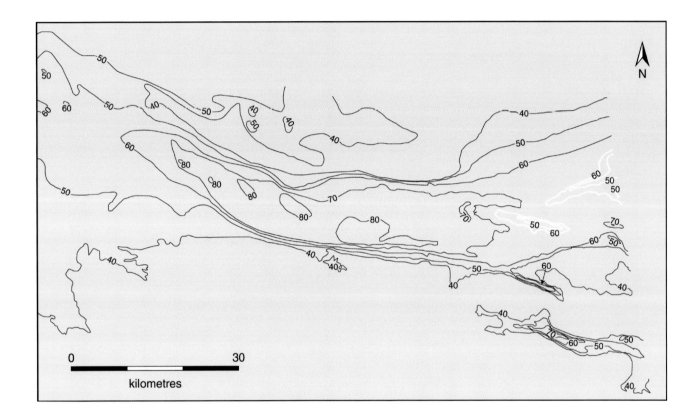

Figure 5.2 Bathymetric contours (40, 50, 60, 70 and 80m below sea level) of the Outer Silver Pit area; depicted in white are the elongate ridge features

5.2 Feature descriptions

Figure 5.3 and Figure 5.4 show the two elongate ridges within the OSP (Figure 5.2). Further details of the locations, dimensions and trends of the ridges are contained in Table 5.1 and Figure 5.2. The ridges are situated at the eastern limit of the OSP (Figure 5.2) and stratigraphically they lie at, or very close to, the seabed, approximately 40-50m below sea level at their crests (Figure 5.3). Ridge A is elongate and tapers towards the south east but is rounded towards the north west end. It is discretely located within the OSP, entirely disconnected from the banks of the depression. Although tapered at its southwestern end, Ridge B broadens to the northeast until it connects to the bank of the OSP at its eastern extremity.

Table 5.1 Position, dimensions and trends of Ridges A and B

	Ridge A	Ridge B
Length	c.18km	c.16.5km
Width	c.2km	c.3km
Height	c.30m	c.20m
Trend	ESE/WNW	WNW/ENE

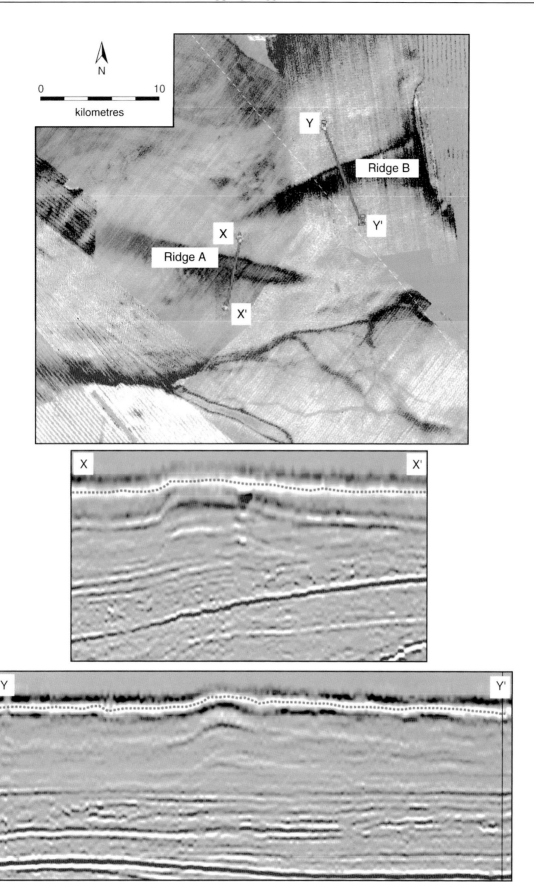

Figure 5.3 Hilbert transform time slice at 0.06 seconds (top) depicting the two ridges. Two arbitrary lines are annotated which pass through Ridges A (X-X) and B (Y-Y).

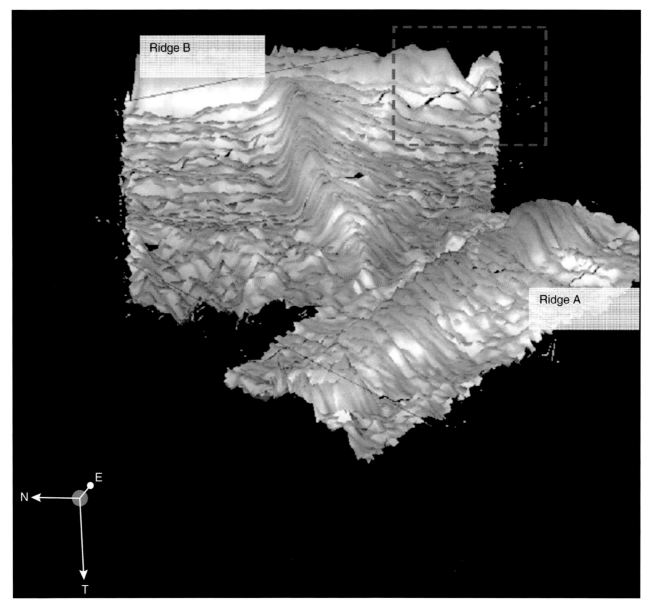

Figure 5.4 3-D illuminated view of Ridges A and B, with a depression referred to in the text highlighted in red. The ridges are vertically exaggerated.

Table 5.2 Average dip of the flanks of Ridges A and B

	Ridge A		Ridge B	
	Shallow Slope	*Steep Slope*	*Shallow Slope*	*Steep Slope*
Major Axis	0.07°	0.1°	0.09°	
Minor Axis	0.035°	0.84°	0.035°	1.13°

In cross section, Ridge A displays an asymmetry along its major axis and both ridges are asymmetric along their minor axes (Figure 5.3 and Figure 5.4); the asymmetry of the minor axis of Ridge A decreases towards its southeasterly end. The steeper slopes of the ridges are slightly concave whereas the shallower slopes are predominantly convex (Figure 5.3). The dip of the ridges' flanks are relatively shallow (Table 5.1 and Figure 5.5), reaching an approximate maximum of only 1.13° on the steeper slope of Ridge B. Given that the trends of the ridges are offset the steeper slope of Ridge A faces south east whereas the steep flank of Ridge B faces south west (Figure 5.3 and Figure 5.4). Cross sectional views of the features (Figure 5.3) also reveal that the crests of the ridges are relatively flat.

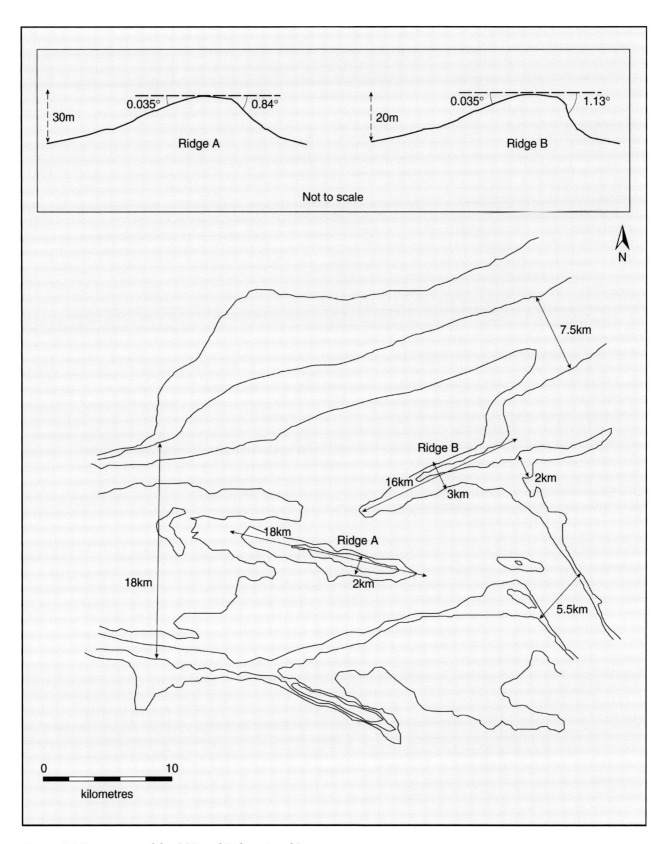

Figure 5.5 Dimensions of the OSP and Ridges A and B

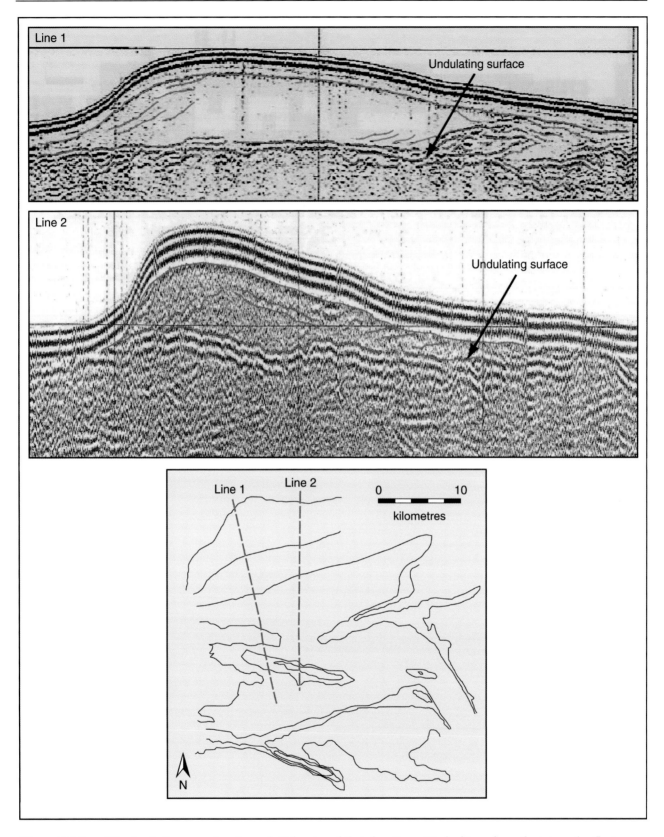

Figure 5.6 Two 2D seismic lines running through Ridge A and their location. Marked in red are the internal reflectors and at the base, the undulating surface upon which the ridge lies

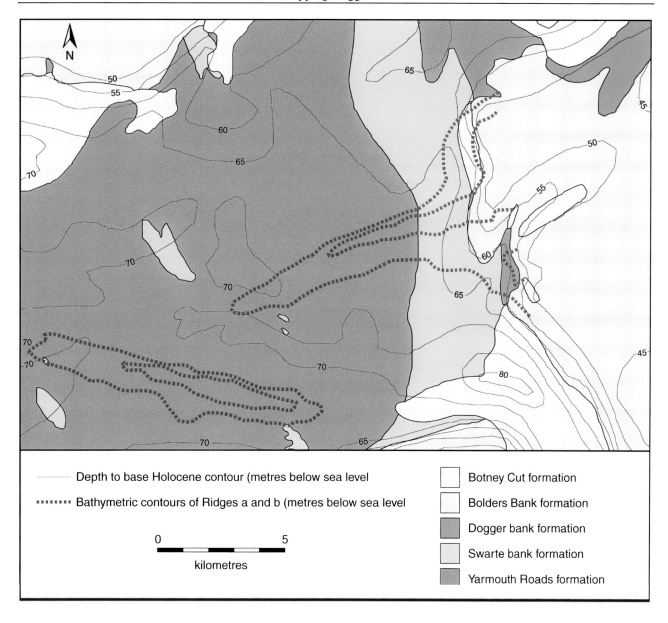

Figure 5.7 Quaternary geology map of the eastern end of the OSP overlain with contours of the depth to the base of the Holocene sediments (black) and the bathymetry of Ridges A and B (dashed red). (Adapted from Larminie, 1989a and 1989b)

Two available 2D seismic lines (Figure 5.6), which pass through Ridge A, provide further details not apparent in the lower resolution 3D data. Firstly, the surface of ridge is smooth on both the steep and shallow slopes. Secondly, the ridge lies on a surface that undulates relatively sharply, truncates the strata below and is seen throughout most of the OSP, where a relatively thin veneer of sediments covers it. Finally, the 2D data shows that the internal structure of Ridge A is composed of a series of dipping internal reflectors (foresets).

On a broader scale, the ridges are located at the confluence of several features. At the eastern end of the OSP, and to the sides of Ridge B, the depression forks into two lesser branches (Figure 5.2, Figure 5.5 and Figure 5.7). The branch to the north of Ridge B trends in an northeasterly direction and is approximately 7.5km wide with a shallow northern flank. The second branch, to the south of Ridge B, trends in a southwesterly direction and is approximately 5.5km wide with relatively steep banks on both sides. It is not possible to constrain the length of the either branch as they continue beyond the confines of the data. Also, to the immediate south of Ridge B lie two lesser depressions measuring approximately 2km and 0.75km in width (Figure 5.4 and Figure 5.7) that emerge from the raised ground to the east of the OSP. The Quaternary geology map of the area reveals that these small depressions correspond with outcrops of the Botney Cut Formation (Figure 5.7), a system of partially or completely infilled subglacial valleys dated to the late Weichselian (Larminie, 1989b).

The geology map of the Holocene and seabed sediments (Figure 5.7) reveals many significant properties of the ridges. The contours of the depth to the base of the Holocene sediments in the Silverwell area shows an absence of the ridges in the topography, thus it can be concluded that the ridges are of Holocene age. The Holocene sediments of the ridges consist of the Tershellingerbank Member (of the Nieuw Zeeland Gronden Formation), which is defined as open marine sediments and are slightly muddy sands (mean grain size 120-300μm) that have been derived from Pleistocene glacial and periglacial deposits (Larminie, 1989a). These sediments are overlain by sands at the seabed and in relation to the surrounding areas this is a larger sediment size than present at the seabed of the OSP depression but of the same size or smaller than the seabed sediments on the nearby higher ground. Furthermore, Figure 5.8, an RMS amplitude map of the ridge area, depicts the sand ridges as having a high amplitude signal commonly associated with sandy sediments.

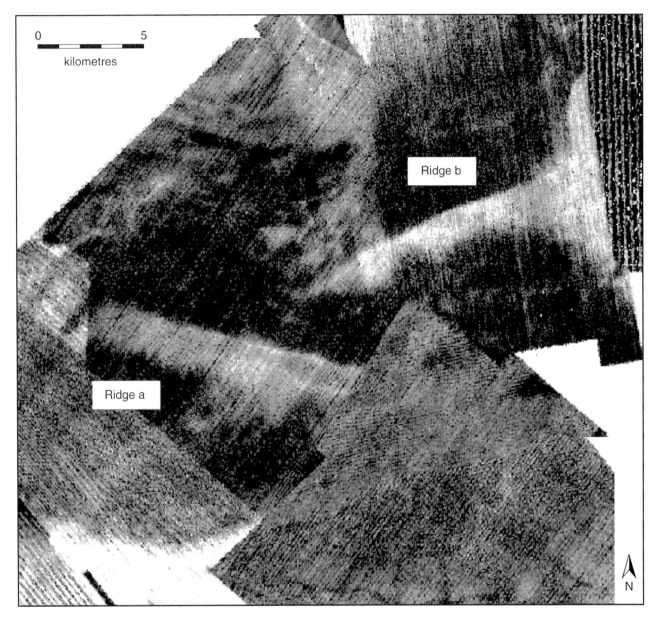

Figure 5.8 An RMS amplitude map from 0.004 to 0.05 seconds (TWT) of the eastern end of the OSP. N.B. the lighter colours depict high amplitude response areas

5.3 Discussion and Feature classification

Deterioration of ice masses following the last glacial maximum has led to a global sea level rise of approximately 120m (Shackleton, 1987) over the past c.20,000 years. In the British Isles a large proportion of the deglaciation had occurred by 13,000 BP (Lowe and Walker, 1997) uncovering vast tracts of the present North Sea Basin as a land surface that was gradually inundated during the Holocene. Shennan *et al.*, (2000) reconstructed the pattern and timing of inundation in the North Sea basin during the Holocene using reliable indicators of past sea levels obtained from sediment core analysis and geophysical models that integrated ice sheet reconstructions, the Earth's rheology, eustasy and glacio- and hydro-isostasy. It was predicted that the OSP had become a shallow estuary by 9,000 BP, the Dogger Bank had become isolated from mainland Europe at high tide by 8,000 BP and by 6,000 BP the Dogger Bank had become completely submerged. Thus, since it is known from the Holocene geological maps and the bathymetry that Ridges A and B were formed at some time during the Holocene (10,000 BP to present), it is suggested that they are the product of either terrestrial or marine processes.

5.3.1 Terrestrial geomorphological processes

During the early Holocene, in the period prior to marine inundation (10,000 BP to 9,000 BP), a number of terrestrial landscape processes common to temperate environments are likely to have sculpted the land surface of the SNS. There are two such classes of processes that could have led to the reworking and deposition of such volumes of sediment retained in Ridges A and B, namely mass movement and fluvial processes. Mass movement, the transfer of material down a slope under the influence of gravity (Summerfield, 1991), is capable of generating significant deposits of fine (colluvium) to large grained (talus) sediment. Inherently, sediments transported by mass movement are deposited at the base of the slope from which they originated. Thus, the situation of Ridge A at 4-5km distance from the proximal slope eliminates the possibility of it being deposited by a gravitational process. Although Ridge B is attached to a slope of the OSP depression, it too can be disregarded as a mass movement deposit as the feature is approximately the same height as the adjacent slope and is thus by no means a mass of material deposited at the base of it.

It is more reasonable to assume that fluvial processes contributed to the shaping of the landscape in the OSP during the very early Holocene. Runoff would have been routed to, and exploited, any local depressions. Thus it is a distinct possibility that Ridges A and B were formed in a fluvial environment. Fluvial bedforms, specifically bars, can achieve significant dimensions, developing lengths that are comparable to the width of the channel in which they form (Knighton, 1998). They can take on a variety of shapes and occur in a range of conditions (Knighton, 1998). The point of branching in the OSP channel at its eastern end (in the vicinity of the two ridges - see feature descriptions), may have hosted a confluence of two rivers under fluvial conditions; several observations exist of bars that have formed at confluences in modern fluvial systems (e.g. Melis *et al.*, 1994; Rhoads and Kentworthy, 1995; De Serres *et al.*, 1999). It is therefore probable that features similar to Ridges A and B would have formed in the position in which they currently exist. However, without going into further details of the morphology and formation of such bars it is possible to discount, with some confidence, Ridges A and B as being fluvial bedforms from the very early Holocene. This is because significant evidence exists that is suggestive of at least one major erosional event in the OSP during the early Holocene.

The Quaternary geology map of the OSP area reveals that, aside from the occasional patches, there is a general absence of late Pleistocene/Late Glacial Maximum sediments in the depression, which makes it somewhat distinctive from the continuous cover in the surrounding area. Although the absence of these sediments cannot be taken as sole evidence of their erosion, it is a notable spatial correlation and reasoning would suggest that it is highly improbable that the ice sheet that covered the OSP area during the Late Glacial Maximum (Carr *et al.*, 2006) would not have deposited material either directly, or in the form of proglacial outwash, in such a selective manner in this region[1]. Evidence of erosion in the OSP during the early Holocene is reported by Larminie (1989a). This indicates that where the Holocene sediments overlie the Botney Cut Formation (a Late Glacial Maximum tunnel valley infill), they are separated by an erosional contact. This is supported by 2D lines crossing the OSP.

In Figure 5.9 a 2D line, situated to the north of Ridges A and B, shows the base Holocene reflector truncates the strata below (the Botney Cut Formation). The sediments recorded as overlying the erosional surface in the OSP are described as fully marine (Larminie, 1989a); consequently, it is possible that marine processes led to the erosion of late Pleistocene sediments and any early Holocene terrestrial sediments that may once have been present in the OSP. It is also likely that the strong erosional processes capable of removing these sediments would also have reworked Ridges A and B had they been present. As the 2D lines in Figure 5.6 show, the ridges overly the truncated surface and thus it is likely that they are a product of marine processes that operated subsequent to the recorded erosional event.

5.3.2 Marine geomorphological processes

The entrainment and subsequent transportation of the large volumes of sand sized sediments that comprise Ridges A and B would have required relatively strong currents over a sustained period of time. At present the OSP is a low energy environment thought to be sheltered from the high tidal current velocities, conveyed from the north, by the shallow Dogger Bank (Eisma, 1975); it is also understood that the OSP is void from the influence of surface waves that are not considered to penetrate beyond 30m in depth in the SNS (McCave, 1971).

---

[1] The Holocene map also shows that the early Holocene, brackish marine sediments of the Elbow Formation, that exist in the SNS and in the locality of the OSP are also absent from the depression. Because the OSP is a closed depression it is possible that there were no Holocene intertidal deposits; when sea level rose sufficiently to cross the threshold of the depression water would have flooded the pit thus the inundation would have been relatively sudden and not a gradual incursion.

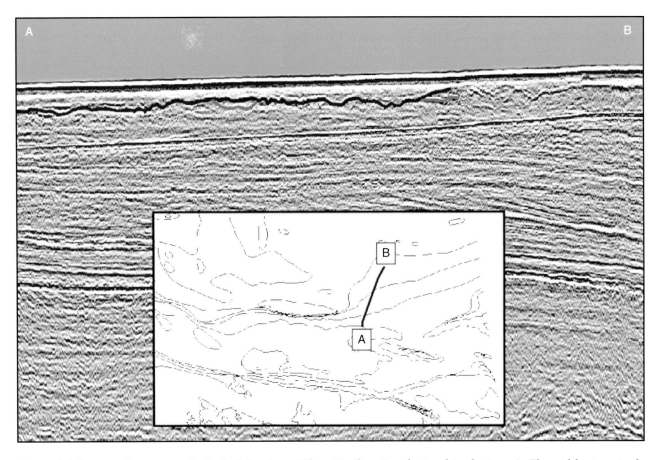

Figure 5.9 Truncated strata on the bed of the Outer Silver Pit (location depicted in the insert). The red horizon is the Holocene Base reflector (Hebbeln and Meggers, 1999); the blue horizons show the strata that have been truncated

Such low hydraulic energy has led to sedimentation on the bed of the OSP[2] of fine sands and silts (Veenstra, 1965; Larminie, 1989a) which are locally laminated[3] (Larminie, 1989a). Indeed, the late Holocene sedimentary record of the OSP is similar to that of the present (Larminie, 1989a) suggesting that the depositional environment was also similarly low energy. Furthermore, Shennan *et al.*, (2000) propose that by 6,000BP the coastline at the margins of the North Sea was comparable to that at present, and therefore it is feasible to suggest that the currents and water depths around the OSP were similar. Consequently, it is likely that Ridges A and B were formed in the early Holocene when the water depth in the OSP was relatively shallow and the currents relatively strong. However, insufficient evidence can be derived from the available sediment and sea level records to further constrain the precise marine influenced environment the ridges were formed in, i.e. estuarine, strait or shallow clastic shelf environment. Therefore, in order to gain such palaeoenvironmental insight the morphology of the features must be examined in respect to modern features of similar form.

The distinctive elongate morphologies of the ridges are similar to a number of landforms that develop subaerially along coasts and on the bed of continental shelves or tidal inlets.

5.3.3 Subaerial coastal landforms

Of the abundance of subaerial coastal landforms that exist globally, Spits and Barrier Islands are identifiable as being depositional elongate ridges. Spits protrude into estuary/bay mouths or the sea whilst attached to the mainland at one end (Masselink and Hughes, 2003). They generally have one or more landward pointing recurves at their distal ends, although they can be entirely linear (Bird, 2000). Spits form when there is a sudden break in the direction of the coastline but sediment transport and deposition continues along the original pathway (Haslett, 2000). They are built up above high tide level and lagoons and marshes develop on the sheltered landward side.

There are several aspects of Ridges A and B and their context which suggest that it is unlikely that they are spits. Al-

---

[2] As mentioned in the previous section the sediments on the top of Ridges A and B are of a larger grain size than those on the bed of the OSP but similar to those at equivalent depths below sea level. This is suggestive of the action of currents and/ or wave action on the tops of the ridges.

[3] The laminations in the OSP form as a consequence of the yearly generation and deterioration of a thermocline at 20-30m depth that is able to develop as a consequence of the minimal impact of tidal action (Eisma, 1975).

though Ridge B protrudes from 'land' at a change in direction of that land body, Ridge A is positioned independently at a point where there is no directional change in the 'coast' and thus suggests that Ridge A was not formed by processes of continued longshore drift of sediment along a pathway not contiguous with the coastline. If these features were spits, it would be likely that the amplitude signal from the 'sheltered' area on the landward side would be noticeably lower as a result of the fine sediments that accumulate in low energy environments of lagoons or marshes.

Figure 5.8 shows that there is no notable difference in the amplitude signal between the land adjacent to the ridges on either side, suggesting the absence of a lagoon or marsh. However, it must be noted that it is possible that these sediments were eroded during the Holocene transgression. The longshore drift processes that are responsible for the formation of spits are likely to be minimal within an estuary or strait environment where the flow dynamics are likely to be dominated by the flood and ebb currents of the tide. It may have been possible for a spit to form at a point during transgression if the easternmost land body in the OSP had been an open coastline. Figure 5.10 however, shows that it was not and that it was inundated at an early point relative to other local land bodies. Finally, although in certain circumstances it is possible that spits can be linear, Ridges A and B lack the recurved end that is frequently characteristic of spits.

Barrier Islands are linear, shore parallel sand bodies that, like spits, extend above sea-level (Masselink and Hughes, 2003). They vary greatly in size and can be up to 100m high, 100s of metres wide and 1000s of metres long (Haslett, 2000). Further similarity to spits extends to the low energy lagoons that are often situated between the barrier and the mainland (Masselink and Hughes, 2003). Again, there are several features of Ridges A and B and their locality that suggests these features are not barrier islands. Firstly, neither of the ridges lies parallel to a potential shoreline. Secondly, at a time when these features would still be in part above sea level, they would have been situated in an inlet (estuary/strait) environment and not on the open coast where Barrier Islands are observed to form (Figure 5.10). Finally, as previously stated there is no suggestion in the amplitude data of the presence of fine lagoonal sediments in a possible lee area of either ridge, thus suggesting the (possible) absence of the sheltered environment that exists between barrier islands and the mainland today.

5.3.4 Marine bedforms

The possibility that Ridges A and B are either submerged Spits or Barrier Islands is rejected and alternatively a suite of marine bedforms are considered. There are three main elongate sandy bedforms that are commonly identified in marine or tidally influenced environments. These are sand ribbons, sand waves and sand banks/ridges. Sand ribbons are longitudinal bedforms that develop parallel or sub-parallel to the dominant tidal flow current (Cameron *et al.*, 1992). They occur in areas with relatively high surface current velocities, often in excess of 100cm/s (Johnson and Baldwin, 1996). Although sand ribbons vary greatly in size they are generally less than 15km in length, 200m wide and up to 1m thick (Kenyon, 1970), however, as a result of their diminutive thickness they tend not to be identifiable on seismic images.

Sand waves, small/medium subaqueous dunes or megaripples as they are sometimes referred to, are flow transverse features, i.e. their crests lie approximately perpendicular to the direction of the main current (Cameron *et al.*, 1992; Blondeaux, 2001). They are greater in height than sand ribbons, generally falling between 1.5m and 10m in thickness (Ashley, 1990; Johnson and Baldwin, 1996) and they generally occur in areas where current velocities exceed 65cm/s (Johnson and Baldwin, 1996). Sand waves can be both symmetrical and asymmetrical in form (Blondeaux, 2001) with maximum slope angles of 10-12° (Cameron *et al.*, 1992).

Ridges A and B are both significantly larger (Table 5.1) than the reported dimensions of sand ribbons or sand waves and thus it is concluded that they are neither feature. Conversely, sand banks/ ridges are much larger features commonly measuring up to 80km long, 1-3km wide and 10-50m high (Johnson and Baldwin, 1996; Dyer and Huntley, 1999) and are of similar dimensions to Ridges A and B. Furthermore, in agreement with the morphology of Ridges A and B, Johnson and Baldwin (1996) describe sand banks/ ridges as linear bedforms that are asymmetrical in cross section and composed of medium to fine sands.

The morphological evidence is strongly suggestive that Ridges A and B are indeed sand banks/ridges in terms of both the similarity which exists between them and also the elimination of a range of other possible features based on the available data. However, this study is lacking in the provision of data that may provide validation for such claims.

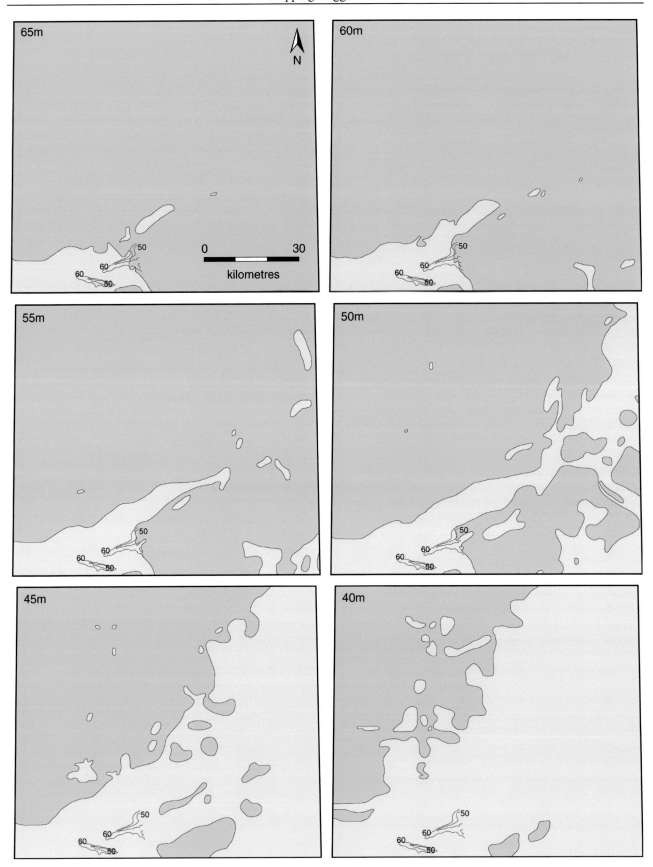

Figure 5.10 Changes in land area with rising sea level based upon the depth to base Holocene map (Larminie, 1989b). Ridges A and B are depicted in red

For example, higher resolution seismic data and sediment cores may be able to provide greater detail of the internal structure of the features and thus reveal the presence/absence of certain sedimentary structures characteristic of sand banks/ ridges[4]. Also access to high-density sediment analysis would provide information of the existence/absence of any sediment grain size gradations that occur across the features[5]. However, the body of evidence available is considered to be sufficient to classify, albeit tentatively, the features in the OSP as sand banks/ridges. Thus the following inferences based on this classification, are made with a degree of uncertainty.

It is possible to make the distinction between sand ridges and sand banks based upon the criteria that ridges have a length/width ratio that exceeds 40 (Amos and King, 1984). The length/width ratios of Ridges A and B are just 9 and 5.5 respectively and so are herein referred to as 'sand banks'.

Sand banks may exist in either an active or moribund state. Active sand banks are present in areas where the tidal currents are relatively strong (>50cm/s). Their crests are shallow in the water and are generally quite sharp, except when they approach sea level at which point they display flattened tops. The steep slopes of active ridges are relatively steeply inclined at c.6° and they are flanked with sand waves. Conversely, moribund sand banks are situated in relatively deep water where currents have diminished to <50cm/s and are thus insufficient to transport sand on the seabed (Johnson and Baldwin, 1996; Dyer and Huntley, 1999). They have rounded profiles, shallow slopes of <1° and an absence of sand waves on their flanks (Johnson and Baldwin, 1996; Dyer and Huntley, 1999). The surrounding sea floor is covered by sandy or muddy sediments as opposed to larger gravel sized sediments found in the vicinity of active banks (Johnson and Baldwin, 1996; Dyer and Huntley, 1999). As previously stated it is known that the currents in the OSP are at present very weak, the crests of the sand banks are flattened and at a depth of approximately 40-50m below sea level and the steep slopes have mean angles of only 0.84° and 1.13°. The 2D seismic data depicts the surfaces of the sand banks as smooth and thus void of sand waves. Thus the sandbanks in the OSP are interpreted as being moribund on the basis of the geomorphological observations.

The moribund sand banks within the OSP are thus relict features formed in conditions that no longer exist at this location. Such features can provide insight into past environments through the understanding of processes that operate to form analogous landforms in the present day. Thus it is the purpose of the following section to utilise the available data in an attempt to reconstruct the processes that operated in the OSP when the sand banks were last active, i.e. the Early Holocene[6]. Due to the rapid and significant extent of sea level rise in the North Sea during the early Holocene, marine conditions within the OSP would have undergone significant transformations e.g. from estuarine, to strait, to open sea. Therefore the suite of processes which initially led to the formation of the sand banks may have been different to those that ultimately shaped and maintained them prior to their becoming moribund.

5.4 Environmental Interpretation

5.4.1 Sand Bank classification

Sand banks develop in a number of marine settings where there is an abundant supply of sand and currents strong enough to transport the material. Dyer and Huntley (1999) produced a classification system of the various types of sand banks resulting in their following subdivision:

i) offshore banks,
ii) estuary mouth banks (including those formed on tidal deltas and those formed in wide estuary mouths where there is not a delta) and
iii) headland associated banks (separating those banks formed around stable and recessional headlands).

Offshore sand banks are described by Blondeaux (2001) as rhythmical features that generally have a crest spacing of a few kilometres. They form with their long axes orientated at a small oblique angle to the peak tidal flow direction, where there is a convergence in bed load transport paths (Dyer and Huntley, 1999). The crests of the banks are often at or only a few metres below the sea surface at low tide. Although there is more than one 'ridge' within the OSP, sets of offshore sand banks/ridges that have been identified in the SNS commonly occur in groups of ten or more (Cameron et al., 1992). Furthermore at a point when the crests of the banks would have been at or close to sea level at low tide (50-40m below present) the OSP would have been an enclosed seaway (i.e. an estuary or a strait)[7] not an offshore environment. Thus it is concluded that the sand banks within the OSP cannot be classified as 'offshore banks'.

It is also possible to conclude that Ridges A and B are not headland associated banks. The crests of the sand banks in the Outer Silver Pit are at approximately the same depth below sea level (c.50m) as the surrounding 'land/potential headland'. Thus at the point when the water depth in the

---

[4] The internal structure of linear sandbanks within estuaries is likely to vary according to the precise controls upon its formation and maintenance. However, one may expect to observe such structures as clay drapes, rippled beds, 'ebb foreset packages' (Fenies et al., 1999).
[5] Larger particle sizes are likely to be observed on the flank where the dominant currents operated.

[6] As previously noted, by 6,000 BP the North Sea sea levels, and consequently the conditions in the OSP, were similar to those at present.
[7] Assuming a tidal range of less than 10 metres based on the current (~8m) and Holocene tidal ranges (<8m) of the Humber estuary on the east coast of England (Shennan and Horton, 2002). Data for Holocene tidal ranges of the OSP are not currently available.

OSP was sufficient to submerge the sandbanks, the adjacent 'land' would also have been submerged and thus could not have been a headland. However, it is possible that at some time in their development the ridges were a lesser height in relation to the 'land' but subsequently gained height during inundation and landform evolution. Nonetheless, regardless of the stage of inundation, at a time when these sandbanks could have been active, the OSP would have been a seaway, not an open coastline with significant longshore processes to generate headland banks. Furthermore, several components remain of the sandbanks' morphology that suggests they are not headland-associated banks. Headland associated banks occur in zones of littoral sediment transport convergence where there is an acute change in the direction of the coastline (Dyer and Huntley, 1999). Sandbanks form on one or both sides of the headland and are separated from the headland by a deep narrow channel; this precludes Ridge B from being a headland-associated ridge as it protrudes from the cusp of the 'land' to which it is attached. Headland associated banks are generally only a few kilometres in length and thus are much shorter than the sand banks in the OSP (Table 5.1). The sand banks in the OSP also do not portray the pear shaped form commonly associated with headland banks (Dyer and Huntley, 1999).

On the premise that, when last active, the sand banks within the OSP would have been close to the water surface or exposed at low tide it is possible to infer, based on the submerged topographical data, that the sand banks were probably formed in a strait or estuary environment. Figure 5.11 is a generalised depiction of the possible land surface, intertidal areas and inundated areas using bathymetric data and based on a series of assumptions that, at present, cannot be further constrained:

1. It is assumed that the crests of the sand banks within the OSP were at, what is now, 50m below sea level when they were last active. However, this depth is derived from the bathymetric data which has a 10m contour interval and so theoretically the crest of the bank could lie at a lesser depth, up to c.41m below present sea level. Also, the height of the sand bank may have changed from when it was last active as a result of post inundation deposition. Larminie (1989a) suggests a generalised range of only 1-5m of Holocene sedimentation in the locality of the sand banks. It is not thought that erosion would have had a significant impact upon the elevation of the sand banks since becoming moribund, because for them to be moribund the local currents operating must be so weak as to not transport sand sized material.

2. It is assumed that the topographic expression of the seabed surface represented in the bathymetric data approximates to that of the land surface at the time the sand banks were last active. For the same reasons as stated above in point 1, this may not have been so. The topographic data in the depth to the base of the Holocene map may have presented a better representation of this land surface[8]. However, because the sand banks are Holocene features they are not depicted. In using the bathymetric data it is argued that a more realistic comparison of the sand banks' elevation in relation to that of the surrounding land is acquired as late Holocene deposition is compensated for; this is of course making the further assumption that the sedimentation rate was spatially continuous.

3. The tidal range is assumed to be 10m, with low tide being at, what is today, 50m below sea level. The level of low tide is based on the further assumption that the crest of the active sand banks became exposed at low tide. This is a somewhat robust assumption based on observations of present day sand banks in estuaries (e.g. Wright, et al., 1975). However, in constraining low tide to 50m, i.e. the height of the crest, all of the above errors outlined in point 1 are subsumed. The assumption of a 10m tidal range is an unavoidable gross assumption, as there is no data available for Holocene tidal ranges/prisms within the OSP. Nonetheless, the assumption is founded in the knowledge of present day tidal ranges of the nearby east coast of England and the current understanding of the effects of such a land configuration on tidal prisms. The exact range of 10m was taken for the purpose of simplicity in the production of Figure 5.11 as the contour spacing of the bathymetry is 10m. This assumption is very limiting for this study, as the land surrounding the OSP is very low relief and therefore a small difference in the tidal range could produce very different marine conditions from high to low tide.

5.4.2 Estuaries and Sand Bank Formation

Estuaries are varied and complex environments; the unique suite of processes that operate in a given estuary are a function of several individual factors (e.g. tides, waves, fluvial input and morphology), which are temporally dynamic over the short- and long-term. In order to gain insight and aid interpretation of estuaries and their many facets, several classification schemes of various estuarine attributes have been proposed. Such classifications are also valuable in the interpretation of past estuarine conditions based on relict features, thus potentially aiding an interpretation of past conditions within the OSP founded on the available information of the moribund banks and their surroundings.

---

[8] Although, problems exist with quantifying and identifying the spatial and temporal distribution of erosion, and subsequently inferring the effects the change in topography may have had on hydraulic processes relative to the stages of sandbank formation.

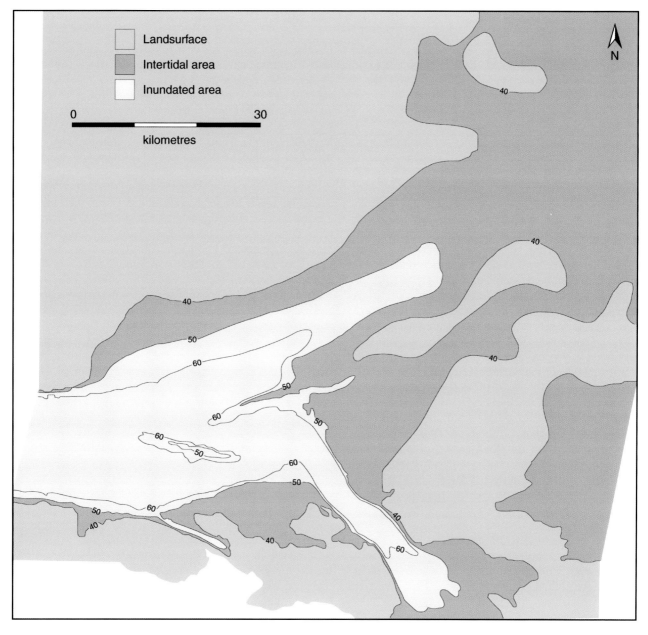

Figure 5.11 Postulated land configuration at the time when the OSP sand banks were last active. The contours are the bathymetric contours from Larminie (1989a). The map is based on a number of assumptions highlighted in the text

Hayes (1975) proposed that estuaries could be grouped according to tidal range. He applied Davies' (1964) scheme of tidal classification that identified three classes of tidal range, micro-tidal (0-2 m), meso-tidal (2-4m) and macro-tidal (>4m). Based on extensive observations of shorelines from around the world, Hayes (1964) concluded that distinctions could be made between tidal ranges and their associated suite of depositional landforms. Most notably he observed that linear sand banks were widely associated with macrotidal estuaries. The currents in macro-tidal estuaries are strong and capable of transporting the relatively coarse grain sediments that form sand banks.

In macro-tidal/tidally dominated[9] estuaries independent and mutually evasive flood- and ebb-dominated channels are widely observed (e.g. Robinson, 1960; Price, 1963; Ludwick, 1975; Wright *et al.*, 1975). These mutually evasive channels develop as a result of the tidal asymmetry that occurs in such estuaries. Tidal asymmetry refers to the asymmetry in magnitude, velocity and duration between the flood and ebb tides in a given estuary (Masselink and Hughes, 2003). The asymmetry is generated as a result of differences in the magnitude of influence exerted by friction upon the flow of each tide. In flow contained within the channel a lesser proportion of the volume of the flood tide (i.e. the crest of the tidal

---

[9] As a consequence of large tidal ranges, the driving morphological processes in such estuaries are tidally dominated.

wave) is in contact with the channel surface (and therefore friction) than that of the shallow ebb tide. Thus the velocity of the channel flow of the flood tide is greater than that in the ebb. Given that the discharge volume through the channel of the flood tide is similar to that of the ebb tide, it follows that the duration of the ebb tide is greater in order to compensate for the reduction in velocity.

In reality the tides in many estuaries are not fully contained within the main channel along its entire length or throughout the full tidal cycle and, frequently, broad zones of intertidal land flank the channel. Friedrichs and Aubrey (1988) examined the relationship between channel shape and tidal asymmetry. They suggested that where large intertidal areas become submerged during the flood tide, the proportion of the water volume exposed to surface friction effects, and indeed the severity of friction[10], is increased and thus the efficiency of the conveyance of the tide up-estuary is reduced. Conversely, during the ebb tide, when water levels are lower, flow is conveyed within the channel where there is a lesser frictional effect. Consequently, the ebb tide has a greater velocity than the flood tide in areas with significant tracts of intertidal land, leading to an overall ebb-dominance. In areas of ebb-dominance, ebb-dominated currents/channels exist within the main channel and the flood dominant currents prevail over the shallow, intertidal areas. To the contrary, in areas of flood dominance, it is the flood currents that dominate in the main channel and the ebb currents that are prevalent along the shallow margins of the channel. It is not uncommon for estuaries to exhibit portions of both flood- and ebb-dominance as a result of the changing morphology along their lengths. Ebb-dominance more frequently occurs towards the head of estuaries where there are larger intertidal zones and flood-dominance mostly occurs towards the mouth. At a given point these regions will interdigitate (Harris, 1988) and, as previously mentioned, the flood and ebb channels tend to be mutually evasive.

It is between the mutually evasive ebb and flood channels within estuaries, where bed loads converge and sand banks occur (e.g. Wright *et al.*, 1975; Harris, 1988; Harris *et al.* 1992; Dyer and Huntley, 1999). The sand waves that exist on the sides of sand banks are commonly observed to be ebb or flood orientated on opposite sides (e.g. Wright *et al.*, 1975; Harris, 1988). Harris (1988) suggests that this is indicative of a 'circulatory pattern around the sandbank crest'. In most cases the tidal flow in one direction will dominate; it is this that causes the sand bank cross-sectional asymmetry (Kenyon *et al.*, 1981). In estuaries sand banks migrate away from their steep slopes, as it is the steep slope that is actively eroded (Dyer and Huntley, 1999). Also sand banks are commonly orientated obliquely to the direction of peak tidal flow (Kenyon *et al.*, 1981).

Only a restricted volume of quantitative studies on sand banks in estuaries exists. Such quantitative study would broaden and improve the depth of understanding of the processes operating to build and maintain the banks. This therefore limits the level of environmental interpretation that can be derived from the sand banks within the OSP. Nevertheless, it is possible to derive some basic understanding of the processes in the OSP during the early Holocene based on the sand banks.

The presence of the sand banks suggests that the OSP was macrotidal and therefore tidally dominated. The tidal currents would have been strong and capable of transporting the sand sized sediments in the sand banks. Tidal asymmetry is likely to have led to the formation of ebb- and flood-dominated channels between which the sand banks may have formed. It is possible that 'Ridge A' was formed in such a way, as it lies independently in the 'main channel' and trends in such a manner relative to the surrounding topography, that it is plausible to suggest it would have formed at an oblique angle relative to the direction of the dominant current. However, the positioning of 'Ridge B', attached to a mass of land and at the confluence of the branch in the OSP depression, may suggest an alternative mode of formation. The convergence or divergence of water moving downstream or upstream along the branches of the OSP most probably resulted in certain hydraulic processes that would have led to a loss of fluid energy and the subsequent deposition of the sand bank sediments. Furthermore, the two smaller depressions directly to the south of Ridge B (see feature descriptions) may have accommodated fluvial and/or tidal channels that are also likely to have influenced the hydrodynamics controlling the processes forming/maintaining the sandbanks. However, it is beyond the scope of this study to examine in further detail the precise fluid mechanics, which may have operated in this area, which resulted in the deposition of 'Ridge B'. It is however, important to suggest that sedimentological analysis and 2D seismic data depicting the internal structure of Ridge B has the potential to provide clarification of the precise processes that operated to form the sandbanks.

5.5 Conclusions

The two elongate ridge features identified in the OSP in the 3D seismic data are interpreted as moribund sand banks that formed in as estuarine environment during the early Holocene marine transgression between approximately 10,000 and 6,000 BP. From this interpretation and knowledge of modern analogues, it is inferred that the tidal range of the OSP was probably macro-tidal and the tidal currents conveyed in the estuary were relatively strong. The identification of a strongly undulating truncation surface in the 2D seismic data suggestive of a major erosional event is supportive of the theories of Donovan (1965; 1975) which suggest that strong marine currents were, at least, in part responsible in the formation of the OSP depression.

---

[10] The roughness coefficient (measure of friction) is generally greater over intertidal areas than within channel as a result of the vegetation that colonises the periodically emergent land.

It is of paramount importance to highlight that this study and its findings are fundamentally restricted by the limitations of both the available data and the level of current understanding of modern, analogous systems. In terms of the available data, limitations arising from the relatively low resolution of the 3D seismic images were, in part, overcome by use of supplementary 2D seismic data. However, the data quality remained insufficient to examine/identify very fine scale sedimentary structures, for example clay drapes and rippled beds, which have the potential to provide validation of the geomorphological classification of the sand banks. Superior resolution 2D line data (e.g. high frequency sonar source) through the sand banks may have allowed for examination of such structures. However, this data was not available. A lack of physical evidence in the way of sedimentary data from cores was unavailable for this study, indeed such data could have provided the means for ground truthing and validation of the interpretations made by providing, for example, information of grain size distribution and detailed internal structure. Furthermore, sediment cores could provide the opportunity for dating and therefore constraint of the timing when the ridges were last active. Finally, current limits in quantitative understanding of the formation and dynamics of sand banks in estuaries, as previously mentioned, render the prediction of the processes that led to the formation of the sandbanks in the OSP problematic. As Knighton (1998) suggests 'ancient deposits and bed forms are a key element in palaeohydraulic reconstruction'; however, 'the reliability of such reconstructions depends on an adequate understanding of the formative processes operating within the present-day environment'.

Despite the limitations and restrictions faced by this study it has highlighted the suitability of 3D seismic data for the identification and broad interpretation of submerged, large scale, geomorphological features and their surroundings. Furthermore, this study has reflected the vital and significant role that relict landforms play in the reconstruction of past environments and specifically palaeohydraulic processes, whilst fervently asserting the importance of the reliability and wealth of data in producing reconstructions that are robust and of scientific value.

6 Salt tectonics in the Southern North Sea: controls on late Pleistocene-Holocene geomorphology.

Simon Holford, Ken Thomson and Vincent Gaffney

6.1 Introduction

During the Upper Permian (*c.* 260 to 251 million years BP) over 1000 m of marine evaporites (the Zechstein Supergroup) accumulated in the North Sea (Cameron et al. 1992). Their subsequent burial promoted mobility on geological timescales resulting in thickness variations from less than 50 m, in regions of salt withdrawal, to more than 2500 m in some of the major salt diapirs (Cameron et al. 1992). Today, the deformed salt deposits encompass a wide range of structural morphologies (Jenyon 1986) and, in places, the crests of salt structures are within 100 m of the seabed (Cameron *et al.* 1992). This proximity to the present-day depositional surface suggests that the uplift and penetration of the overburden by Upper Permian evaporites may well have influenced the topography and hence depositional systems in this region in the recent geological past.

Measured uplift rates of emergent and immediately subsurface salt diapirs range from 2 to 7 mm yr^{-1} (Bruthans *et al.* 2006). This invariably leads to the deformation of the overlying rock layers and hence exerts an important control on synkinematic sedimentation patterns, and hence geomorphic processes. Topographic relief produced when salt approaches the land surface in continental settings can vary from 45 m (Al Salif, Yemen; Davison *et al.* 1996) to up to 1500 m (Zagros Mountains, Iran; Talbot and Alavi 1996), whilst in offshore settings e.g. the Mississippi Delta; topographic relief varies between 100 and 240 m (Jackson *et al.* 1994).

However, a more appropriate analogy for the influence of salt tectonics on landscape evolution in the SNS during the late Pleistocene-Holocene is that of the Five Islands, south central Louisiana. Located on a low-relief landscape near the western boundary of the Mississippi River delta plain, the Five Islands comprise five salt domes aligned in an approximately north west-south east trend, which have pierced and uplifted overlying late Pleistocene meander belt deposits (Autin 2002). The domes are all nearly circular in plan, surrounded by lowland Pleistocene and/or Holocene delta plain marshes, and attain maximum elevations which range from *c.* 23 m asl on Jefferson Island to *c.* 52 m asl on Weeks Island (Autin 2002). The geomorphic impact of the salt domes is most clearly exemplified by Avery Island, where proximal fluvial channels, although modified by engineering in places, show an overall sub-concentric pattern, encircling the salt dome and following the topography closely.

6.2 Relationships between salt structures and late Pleistocene-Holocene fluvial systems.

By comparison with the adjacent onshore landscapes of East Anglia and Continental Europe (Belgium, Denmark) which are characterised by similar subsurface geology, the late Pleistocene-Holocene landscape of the SNS is likely to have been defined by a low-relief, relatively flat land surface. Once Holocene modifications due to sediment accretion (e.g. sandbanks; Stride *et al.* 1982) or erosion (e.g. by tidal scouring) are factored into consideration, the present-day bathymetry of the SNS reveals a relatively flat surface with water depths of only 20 to 40 m in the study area (Cameron *et al.* 1992). It is difficult to reconcile the present-day bathymetry to any distinct structural control by salt tectonics, but it is likely that the subaerial late Pleistocene-early Holocene landscape may well have been influenced by near-surface salt bodies e.g. in the form of relative topographic highs above salt cored anticlines. There is evidence to support this hypothesis from 3D seismic data from the north of this study area. Figure 6.1 presents a series of timeslices (amplitude, Hilbert transform and phase, 0.1 seconds) from the south of the project area. Near the centre of the timeslice a series of broadly concentric reflectors, diagnostic of a salt diapir (Stewart 1999) are clearly visible (Figure 6.1b). An arbitrary seismic line through the 3D seismic volume (Figure 6.1e) confirms the structure is a diapir and hence likely to have been expressed by a relative topographic high on the late Pleistocene-Holocene land surface prior to early Holocene marine transgression. There is some faulting associated with the southern flank of the salt diapir (Figure 6.1b). To the south west of the salt structure an approximately west north west - east south east trending sinuous feature is identified from the seismic time slices (Figure 6.1a-d). This feature is interpreted as a fluvial channel. To the southwest of the presumed fluvial channel there appears to be another, more elongate salt structure. The fluvial channel can therefore be interpreted as occupying a relative topographic low, or even a shallow valley, within the late Pleistocene-Holocene land surface, between two relative topographic highs cored by actively upwelling salt. The topographic low may result from the withdrawal of salt in the subsurface, thereby causing the overburden and land surface to downwarp. Several apparent meander loops can be identified within the fluvial channel, and one meander appears to coincide with one of the radial faults associated with the salt diapir, suggesting a further tectonic control on channel geometry and evolution.

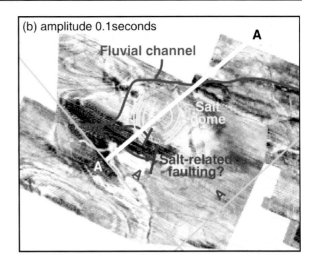
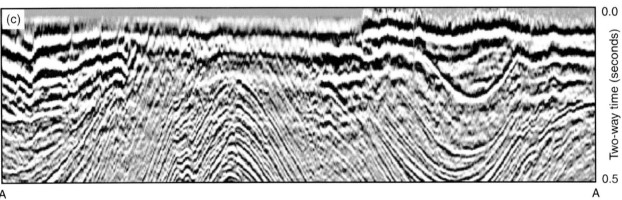

Figure 6.1 (a) Amplitude time slice (0.1 seconds) centred on prominent salt dome with a characteristic concentric reflector pattern. (b) Amplitude time slice (0.1 seconds) with interpretation. A fluvial channel that flows sub-parallel to the salt dome is identified, and it is suggested that salt-related faults have also influenced the direction and geometry of the fluvial channel. (c) Uninterpreted arbitrary seismic line A-A', confirming that the concentric feature identified in (a-d) is a salt dome

Figure 6.2 contains an amplitude timeslice (0.076 seconds) and seismic section. These show an apparently north westerly draining network of channels, which flow towards the eastern end of the OSP. The first-order tributary to this drainage network overlies the axis of a salt-cored anticline (Figure 6.2b). A seismic profile through the channel shows that the channel body directly overlies a possible collapse graben above the salt swell (Figure 6.2c and d). This observation suggests that the formation of the collapse graben led to the development of a topographic depression in the land surface that was exploited by the fluvial channel.

More conclusive evidence for the direct control of collapse graben on late Pleistocene-Holocene fluvial systems is presented in Figure 6.3. This figure contains part of an east-west trending shallow seismic (sparker) profile, 81/03/53, located in the western part of the OSP. Sparker profiles provide higher resolution data compared to offshore 2D and 3D seismic reflection surveys. Immediately below the seafloor reflector is a thin package of Holocene reflections. These are draped over a thicker package of mostly sub parallel Pleistocene reflections, the thickness of which varies laterally. In several places along the line the reflections within the Pleistocene succession terminate against packages of chaotic and complex reflections, which most likely represent infilled tunnel valleys. The base of the Pleistocene succession is marked by an approximately horizontal unconformable surface, beneath which is the steeply dipping, truncated, pre-Pleistocene succession. Here, the pre-Pleistocene succession has been folded above an upwelling body of salt into an asymmetric anticline, the hinge of which has been truncated. Near the truncated hinge of the fold the Pleistocene succession shows a marked thinning. This is indicative of growth of the fold during Pleistocene times (Figure 6.3b). At the hinge itself, the seafloor has a striking horst-graben-horst type morphology (Figure 6.3c and d). This suggests that the overburden above the salt-cored fold has begun to collapse. The fact that the seafloor itself shows such a spectacular collapse, graben-style, suggests that this overburden deformation has occurred within the very recent past, and may indeed be continuing through to the present day.

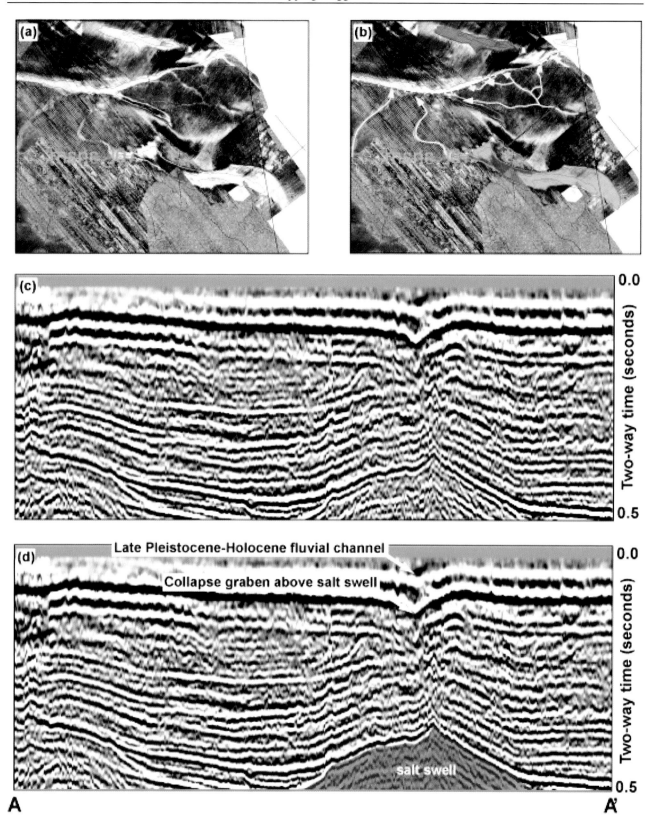

Figure 6.2 (a) Amplitude time slice (0.076 seconds). (b) As (a) but with interpretation; late Pleistocene-Holocene fluvial channels signified by yellow arrows, Holocene sand waves shown in purple, with the green annotation signifying a Pleistocene tunnel valley. (c) Seismic section. (d) Part of seismic section (c) with interpretation

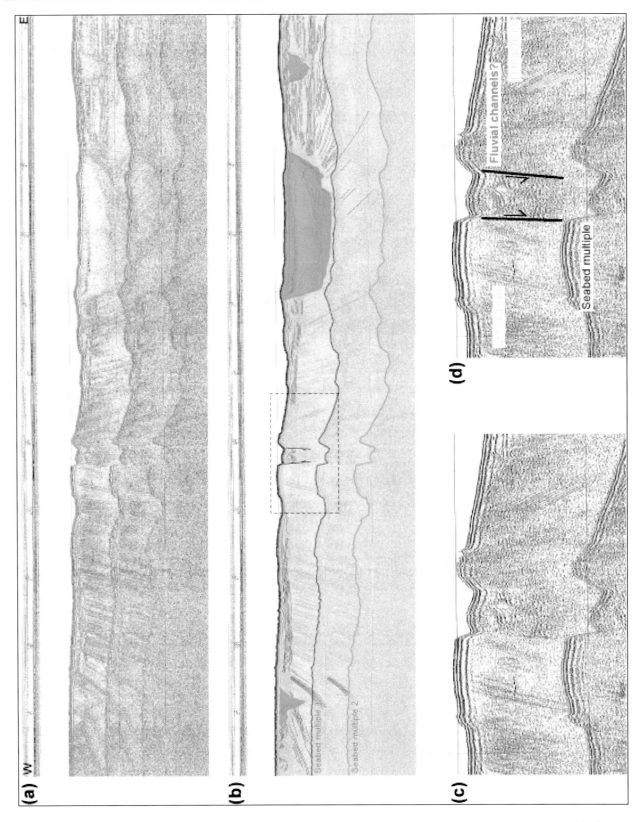

Figure 6.3 (a) Part of BGS sparker profile 81/03/53 that trends east-west through the OSP. (b) Sparker profile 81/03/53 with interpretation. Seafloor reflector and seafloor multiples shown in red. Tunnel valleys shaded in green, Pleistocene sediments in purple and pre-Pleistocene sediments in yellow. Reflections within the Pleistocene succession are picked in dark purple. Within the pre-Pleistocene succession discernable reflections are marked by dark blue picks, whilst diffractions (which mark the radial scattering of incident seismic energy) are shown in light blue. (c) Close up of boxed region indicated in (b). (d) Close up of collapse graben with interpretation, showing channel development within collapse graben.

These observations provide convincing evidence for neotectonic activity driven by salt diapirism. Examination of the Pleistocene succession within the collapse graben reveals a series of characteristic seismic facies units, with sequences of sub-horizontal reflectors downlapping onto inclined reflections; these are interpreted as channel fill deposits. It seems apparent therefore that the collapse graben identified here was active during late Pleistocene times, and moreover, that several fluvial channels were flowing along the axis of the graben.

6.3 Conclusions

Salt tectonics is responsible for some of the late Pleistocene-Holocene geomorphology of the SNS. 2D and 3D seismic reflection datasets provide evidence that salt structure locally influenced late Pleistocene-Holocene drainage patterns, with rivers either flowing around the relative topographic highs above near surface salt domes, or flowing along the topographic depressions created by collapse structures in the immediate overburden above actively upwelling salt structures. A shallow seismic profile from the OSP (81/03/53) provides spectacular evidence that recent salt movement leading to collapse graben formation has not only controlled the location of late Pleistocene-Holocene fluvial systems, but also directly controls the morphology of the seafloor. It is clear therefore that halokinetic activity, which dominates the structural fabric of the SNS basin, has also played a critical role in the late Pleistocene-Holocene geomorphic evolution of this region.

7 An Atlas of the Palaeolandscapes of the Southern North Sea

Simon Fitch, Vincent Gaffney, Kenneth Thomson with Kate Briggs, Mark Bunch and Simon Holford

7.1 Introduction

The preceding papers in this volume have provided the essential background to the results of the NSPP mapping programme. Here, we need only be concerned with a description of the analytical processes that led to the identification of features within the available seismic data and a description and interpretation of these features. A provisional interpretation of the results will be presented in the final chapter.

The analysis presented here utilised the top 0.5s of the SNS Mega Survey provided by PGS for use within the NSPP. The Mega Survey comprises a series of surveys that have been pre-balanced, stacked and migrated into a single contiguous data set. The intrinsic spatial qualities of the data permit the inclusion of supporting data sets to assist interpretation (see Thomson and Gaffney and Fitch et al, this volume). For this purpose, high-resolution boomer surveys acquired for the examination of the Quaternary sediments of the North Sea were sourced from the British Geological Survey (BGS) to assist in the investigation. In addition, other high-resolution seismic information and traditional data sources including core log information were also integrated to provide a framework within which archaeological landscape interpretation could be attempted.

Interpretation of the very large data set used in the project, and the limited period available for study, necessitated the division of data between team members. However, to ensure consistency, interpretative procedures were standardised across the team. This process was facilitated by the use of a consistent colour scheme to represent specific features and the results from each area were appraised by each member of the project team (Figure 7.2).

When a final area interpretation had been achieved, a second evaluation of the interpretation was carried out as a group exercise. Following this, the interpretation was integrated within the project GIS and assessed against supporting data. Geophysical timeslice information was also integrated via a series of pseudo-timeslices generated for specific time intervals. Once created, attribute information for each seismic slice was extracted and applied to the pseudo slice. The resultant information was then extracted as a series of ASCII files containing the necessary X, Y, Z, and attribute information. This data was then used to produce an image of the geophysical timeslice. An RMS Slice was utilised for this purpose, where it was necessary to display an image representing the features within a series of timeslices.

The results of this process for the whole study area are presented in Figure 7.1. For presentation purposes the detail of the results have been presented as four quadrants reflecting the underlying BGS geological maps. When considering these maps, it should be stressed that the primary intent of the atlas is to provide information on features that had a clear physical expression upon the Holocene landscapes under review, in many cases earlier features could be identified.

However, these were not consistently mapped, except on those occasions when earlier features were reflected within the later Holocene landscape. For instance, deeply incised, earlier structures can be demonstrated to form basins for lacustrine features and to constrain later channel development. Despite this, older features which were mapped, but which had no relation to the Holocene landscape, are presented in Figure 7.3 for the sake of completeness.

A further point should be made in relation to topographic variation. Aside from features with clear structural integrity, it was also possible to derive general topographic expression for the wider landscape from the seismic data. This was recorded where observed, and this provides the basis for the general topographic maps for each atlas quadrant and for the final overall map (Figure 7.4). It was not possible, given the time constraints of the project and the extent of the data, to identify extensive surfaces (even where possible) and more detailed information on surface topography. This information must, therefore, be used with care.

Although the results are presented here as a series of paper maps, this is not an adequate reflection of the richness of the data or the utility of the information as a digital model. Recently, the team at Birmingham has been investigating the potential of quadrant Internet mapping and, specifically, Google Earth format to distribute digital mapping to a wider audience (Barratt et al 2007). Aside from archiving with the Archaeology Data Service [1], further information on this and interim data releases may be found on the Project's web site [2].

---

[1] http://ads.ahds.ac.uk/
[2] http://www.iaa.bham.ac.uk/research/fieldwork_research_themes/projects/North_Sea_Palaeolandscapes/index.htm

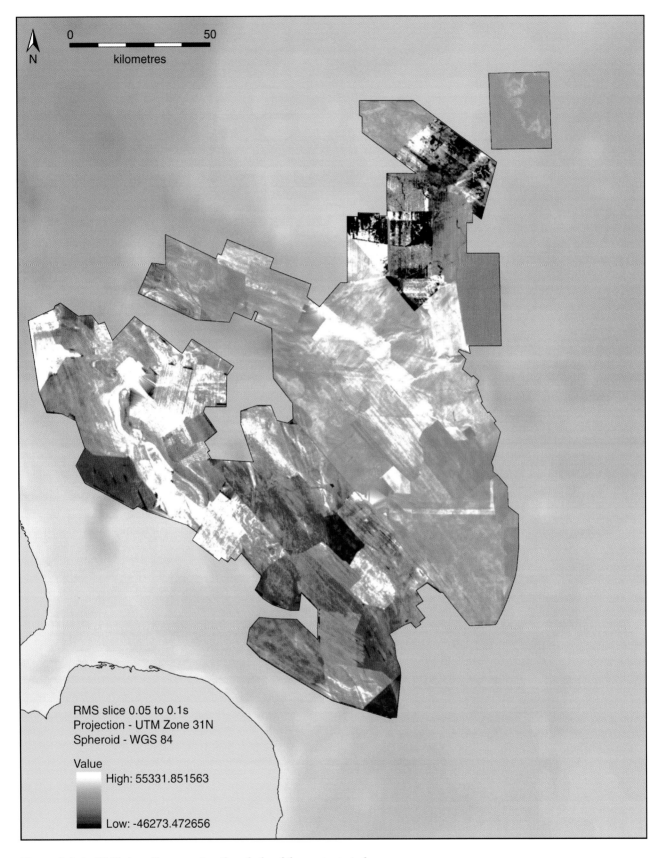

Figure 7.1 An RMS timeslice covering the whole of the project study area

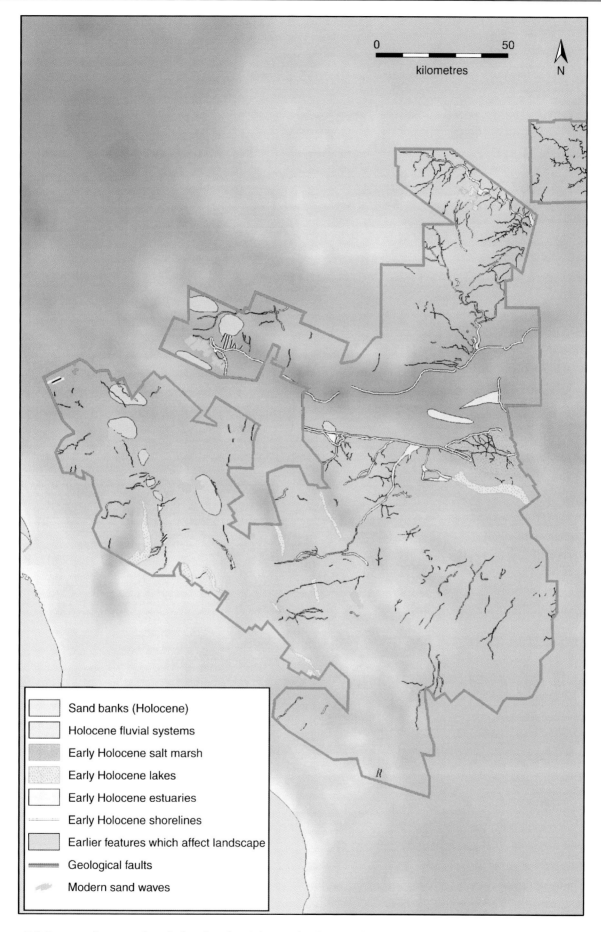

Figure 7.2 Primary features identified within the Holocene landscape of the southern North Sea

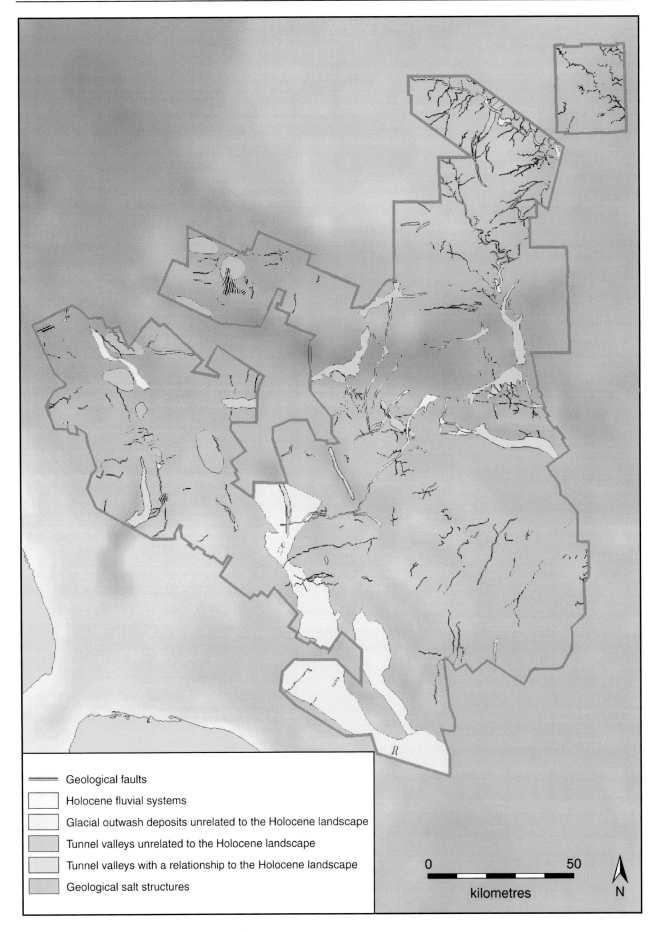

Figure 7.3 Pre-Holocene features recorded during mapping

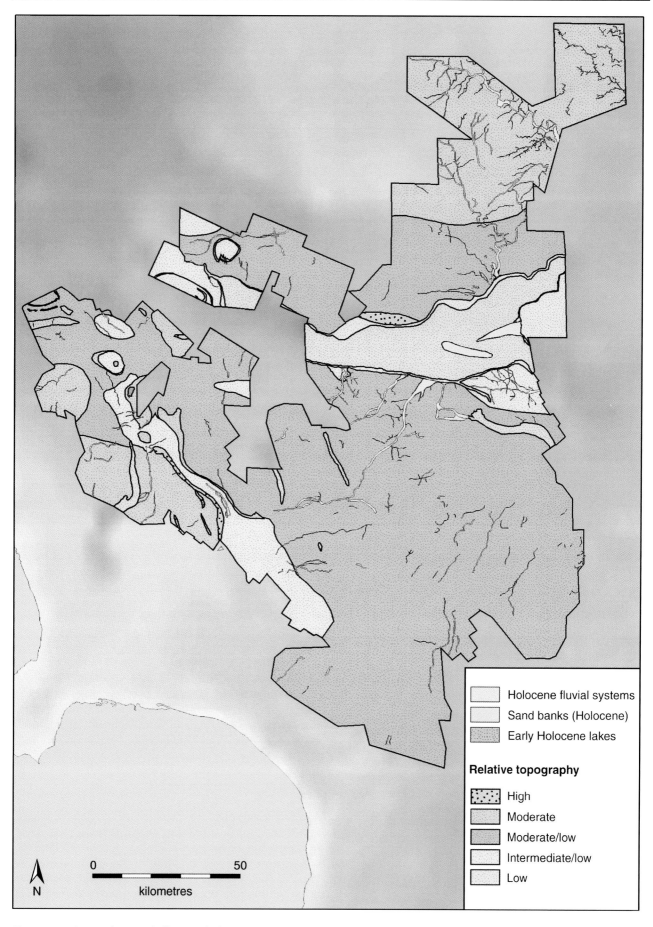

Figure 7.4 General map of all recorded Holocene landscape features including general topographic interpretation

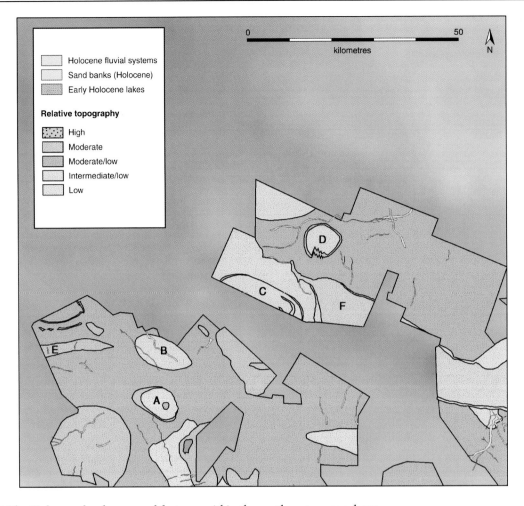

Figure 7.5 The Holocene landscape and features within the northwestern quadrant

7.2 North Western Quadrant

7.2.1 Description

The landscape of the northwestern quadrant displays strong geological influence. This in part is due to the thin Pleistocene sediment cover within the area, which allows underlying geological relief to influence the overlying Holocene landscape (Lumsden 1986a). In addition to this, four active salt domes in this region are associated with graben collapse features and these form dominant structures within the landscape. The first of these circular structures (Figure 7.5 A) appears as a depression in the south of this quadrant. It features an upstanding lip on the west of this structure, and was formed by an outcrop of solid geology that remained upstanding after the collapse of the salt dome graben. In the Holocene, this structure would have created low hills of only a few metres in height. These would have partially surrounded the main graben collapse, which would have created a lower area, possibly containing a marshy area.

The second structure (Figure 7.5 B) appears in close proximity to the previous feature. The expression of this structure is very slight, and it is possible that this had little visible impact on the landscape. This is also suggested by the identification of a fluvial feature that appears to run directly across the feature.

A third structure, (Figure 7.5 C), is located at the mouth of the OSP depression. This structure is surrounded by relatively flat land. However, this graben collapse structure forms a slight, but distinct, depression in this area. The graben collapse is surrounded by slight rises formed by upstanding geology. This area is also likely to have contained a marshy depression during the Mesolithic period, with the upstanding lip forming a slight, but visible, rise in the ground level. The location of this depression close to the edge of the OSP suggests that during inundation, this structure may have bounded a channel joining the OSP to the wider marine environment.

The fourth, and final, structure is smaller than the others. However this crestal collapse differs in being surrounded by a clear ring of upstanding geology (Figure 7.5 D). The outer ring is disturbed in places by geological faulting associated with the main graben collapse. The upstanding nature of this ring is clearly demonstrated by an adjacent fluvial feature that is clearly channelled around the structure (Figure 7.7). This circular structure, however, possesses several interesting properties.

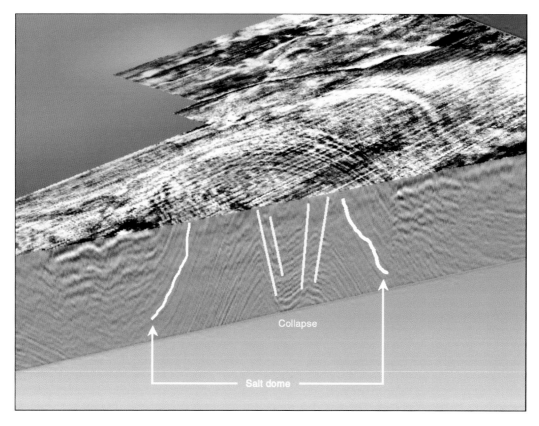

Figure 7.6 Vertical slice through salt dome exhibiting graben collapse

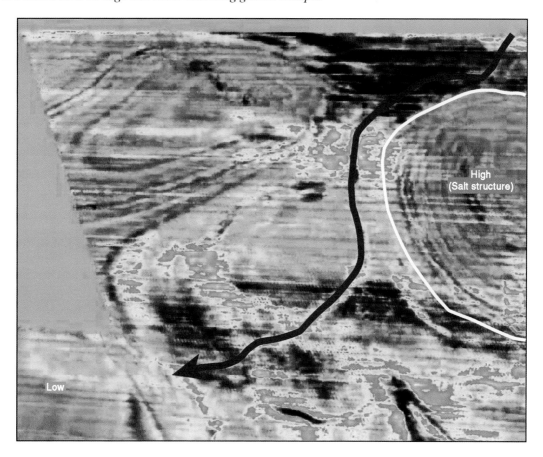

Figure 7.7 The major fluvial channel (blue) can be clearly seen to deviate to respect the topographic rise formed during the Holocene by the underlying salt structure

It may be that the basin is surrounded by solid geology which, given the right conditions, may have retained water. If this is so, two possibilities are suggested. The first is that this structure retained water and formed a lake that was surrounded by low hills. If so this would have formed a highly attractive environment for hunter-gatherers. Another option is that it may have contained a general marshy or wetland area, if the faulting and/or geological permeability prevented significant water build up. Unfortunately, neither scenario can be advanced with certainty. Indeed, as the water table was raised prior to inundation a marshy area may well have become a lake over time.

The fact that these salt structures formed enough of a topographic expression to produce regional highs and lows strongly suggests that at least some salt structures within this region were active during the Late Pleistocene and Early Holocene. This evidence is not inconsistent with observations of recent salt movements in the North Sea (Holford et al. this volume). The large Dogger Bank earthquake recorded in 1931 indicates that the area is still active (BGS 2007).

The underlying geology is also evident in other areas of the quadrant. To the far west, several prominent features may also be observed (Figures 7.5E and 7.8). One is directly correlated with the Flamborough Head disturbance, which appears on BGS mapping as being directly exposed on the seabed. The flanks of this feature are covered by a very thin veneer of Pleistocene material. This would have represented a significant Holocene landscape feature and appeared as a dominating, but low, ridge extending from the present coastline out into the North Sea. Given the thin sediments within this region it is unlikely that extensive archaeological sediments are preserved in the area.

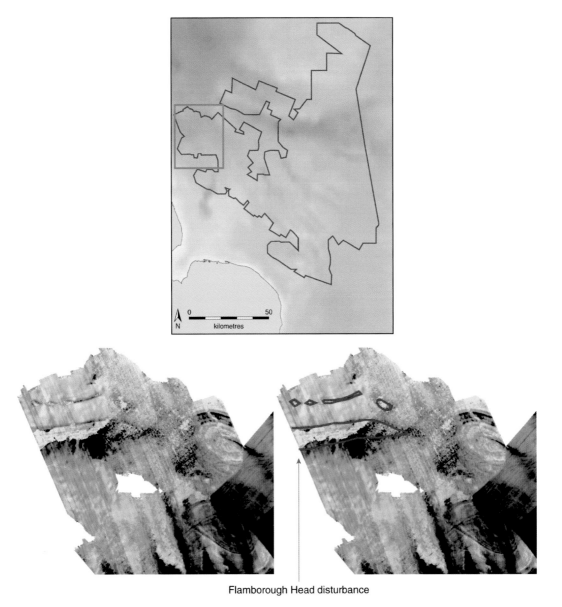

Figure 7.8 Seismic relief image of the Flamborough Head disturbance, note the visible positive relief it imparts to the Holocene landscape

Further south, the Holocene landscape would have risen as it drew nearer to the present shoreline. Within this area a series of small, fragmentary and truncated fluvial features can be observed. Their patchy nature is almost certainly a reflection of post inundation erosion (Flemming 2002, Cameron 1992). The poor resolution of these features may also be due to noise caused by the shallow water column in this area.

The dominant feature in the northeast of this quadrant is the western end of the OSP (Figure 7.5 F and Figure 7.9). A distinct fluvial feature can be seen running west north west from the OSP towards the topographic depression in between Flamborough Head and the Dogger Bank. This feature may be a channel flowing from, or even feeding, the lake that must have filled the OSP during the early Holocene. This channel was certainly active prior to the inundation of this region c. 9.5 Ka BP (Shennan 2000). A similar channel was advocated by Coles (1998) in her map of the region, although at the time no evidence was available to support such a proposal. In any case, the presence of such a channel suggests that the OSP contained a significant freshwater body prior to marine inundation that would have been very attractive to hunter gathers.

With respect of the OSP itself, it is clear that the bulk of this structure has suffered from Early Holocene marine erosion (Figure 7.10). Large scour marks are clearly visible within the depression, and the seismic data clearly shows truncation of deposits. This observation suggests that any lacustrine deposits that remain within this feature are likely to be intermittent. Coastlines at the edges of the Outer Silver Pit are characterised by a strong response and appear as distinct boundaries. Tidal scour marks are also visible and reflect differences in tidal flow. Although no clear dating evidence is available, isostatic models suggest that this coastline was active at around 9,500 BP (Shennan 2000).

7.2.2 Other Features

7.2.2.1 Solid Geology

The underlying bedrock, along with associated faults, is seen clearly within the upper timeslices of the data. Although there are clear indications that the solid geology of this region is near the surface, not all of this need have a topographic expression. Indeed the BGS maps for this region record that late Pleistocene cover in this region is very thin over much of the area (Laraminie 1989). This allows the strongly reflective bedrock to swamp the signal of all but the largest of Holocene features. Thus it is likely that this area may contain features that could not be identified during mapping. This observation must also be tempered by the observation that it is possible that active erosion may have effectively truncated the Holocene deposits and features within in this area. The remaining features in the offshore regions (<-10m) are therefore likely to be incised enhancing their survival and detection.

7.2.2.2 Recent Geological Features

A series of linear features can be discerned in the centre of the study area, near the outflow of the OSP Lake. Their position in the upper sections of the seismic section and their structure revealed that these features were reflections of large sand waves on the seabed surface that are of recent origin (Lumsden 1986b).

7.3 North Eastern Quadrant

7.3.1 Description

The northeastern quadrant of the survey area has provided one of the most complete pictures of the emergent landscape of the SNS. This area was mapped as part of a pilot study prior to this project, and the results of that work have been confirmed and enhanced during this larger exercise (Fitch et. al. 2005). The level nature of this landscape largely reflects the presence of deep Late Pleistocene sediments within the region. The area possesses a topographic high over the area of the Dogger Bank that gently descends into the lower lying plain surrounding the OSP. One minor topographic high can be observed in the southwest of this quadrant, located approximately over the Outer Well Bank (Figure 7.11 B, Laraminie 1989). This topographic high is related to a facies change within the Late Pleistocene deposits, and is due to a change in depositional environment during the Late Pleistocene (Laramine 1989). However the dominant topographic feature in the area remains the *OSP*, which forms a significant depression in the south of the quadrant (Figure 7.11 C).

Within this quadrant the predominant trend of all the fluvial systems is to the southeast. These drain the area of the Dogger Bank, to the north, and converge on the OSP, (Figure 7.11 D, A). All of the Holocene fluvial features can be seen to be incised into the underlying Late Pleistocene Dogger Bank Formation. This relationship demonstrates that these features are likely to have been latest Pleistocene or Early Holocene in date. The majority of these features are highly developed sinuous systems with a high stream order. The geographic location of these systems, in relation to the early Holocene topography, suggests that they were sub-aerially exposed for a longer period than most of the survey area and that the systems are better developed as a consequence. A number of abandoned meanders can be observed, in association with a developed main central channel, and these join the coast via a well-formed estuary. The latter feature can just be seen through heavy striping within the dataset.

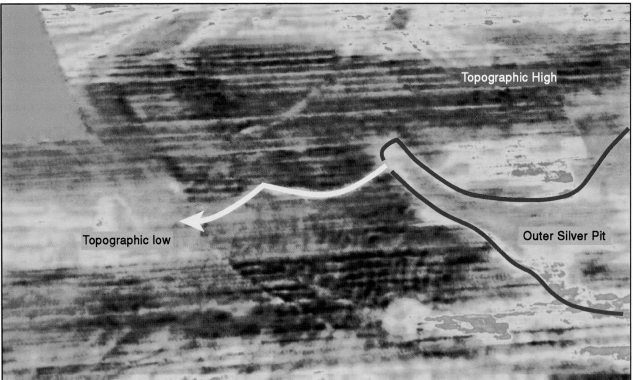

Figure 7.9 Western end of the OSP lake showing outflow channel and associated topographic highs and lows

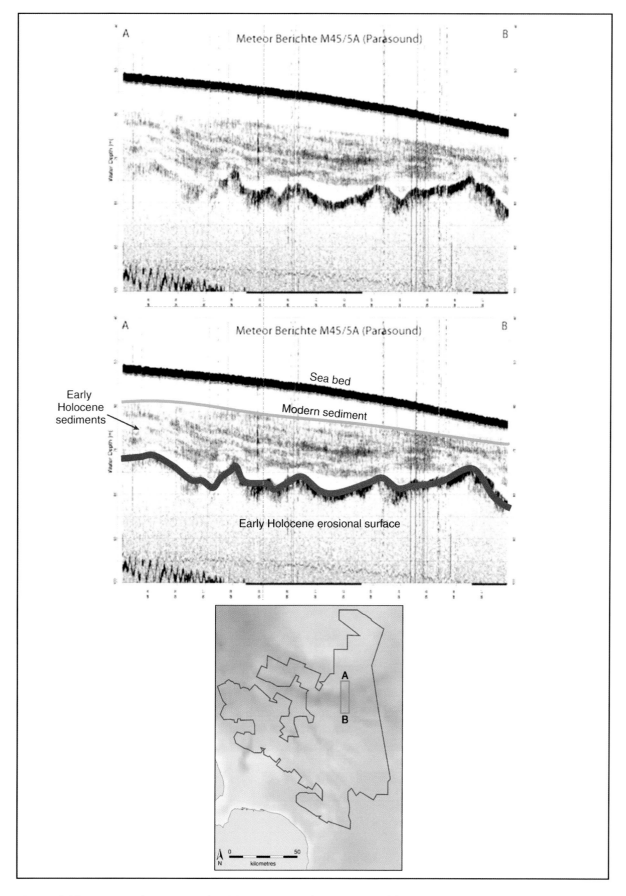

Figure 7.10 A seismic line across the OSP. Pronouced scouring (red line) occured during the early Holocene marine transgression of the area. The surface is overlain by later early Holocene marine sands and muds. (Data provided by the University of Bremen)

The abandonment of channels is presumably a response to changes in the fluvial regime. In this case, the most likely cause is a response to regional sea-level rise. The final landscape features in this quadrant that require description are a series of bulbous basins that occur next to a prominent fluvial system referred to as the "Shotton River" (Figure 7.11 E, Fitch et. al. 2005). Initially recorded during the pilot project these basins have been interpreted as wetlands or lakes. The correlation between these features and an underlying tunnel valley within the seismic dataset suggests that the earlier depression may have been filled with impermeable glacial fill material leading to lake or marsh formation. Again, such wetland systems could have been gathering points for hunter gatherer groups within the landscape and provided a wide variety of hunting and gathering opportunities.

As the fluvial systems progress down to the OSP, clear coastlines are observed framing the OSP itself (between B and D (Figure 7.11). These are characterised by a strong response and are often accompanied on the seaward side by tidal scour marks. A number of rivers meet the coastline in this area, widen and form small estuaries (Figure 7.12). The coast (or lakeside) is clearly defined, located between the -40 metre and -50 metre contours and coincide with the contemporary outline topography of the OSP.

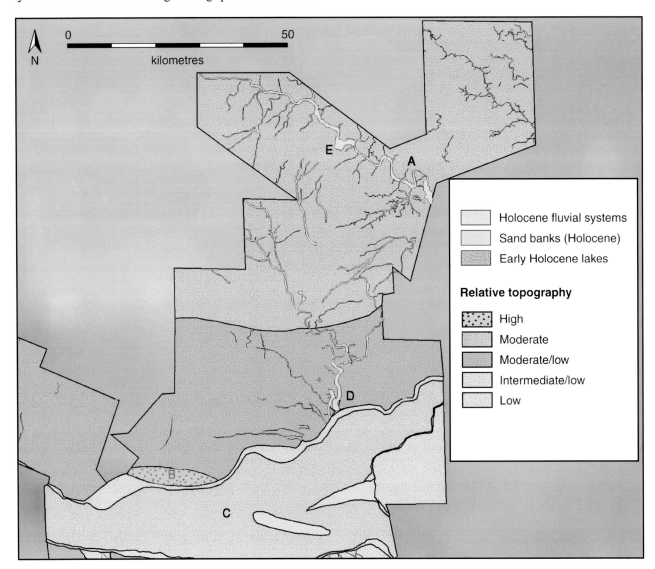

Figure 7.11 The Holocene landscape and features within the northeastern quadrant

If we utilise the sea level curves provided by Jelgersma (1979), Shennan (2000) and Peltier (2004) then an approximate age between 9,500 BP and 8,500 BP is suggested.

Unfortunately, the profusion of sea level curves for this region, combined with the difficulties of relating such models to the real world (Bell et al. 2006), demands that an accurate age for this shoreline awaits the recovery of suitable, dateable material.

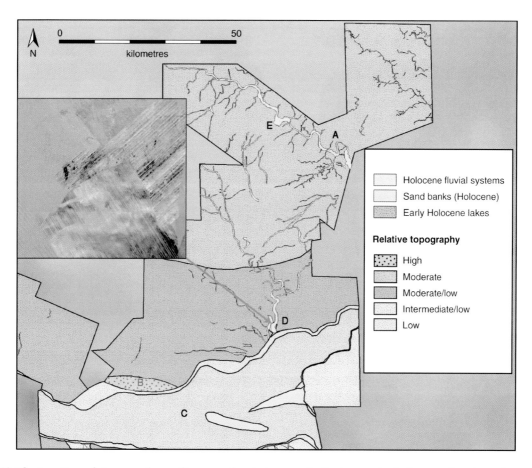

Figure 7.12 The junction of rivers and coastline can clearly be seen in the inset seismic image. The river channels can be observed to widen and form small estuaries

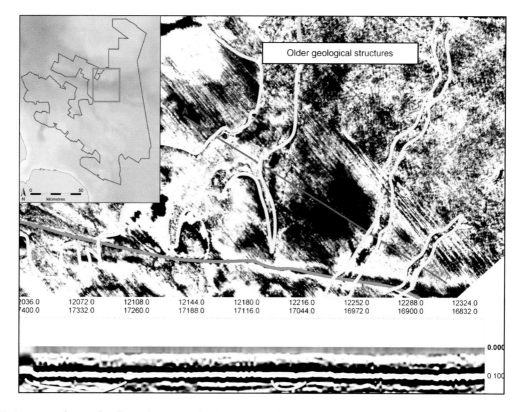

Figure 7.13 A series of tunnel valleys (green outline) crossing the OSP and observed to underlie the Holocene landscape (coastline marked in orange)

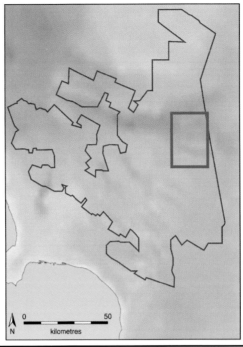

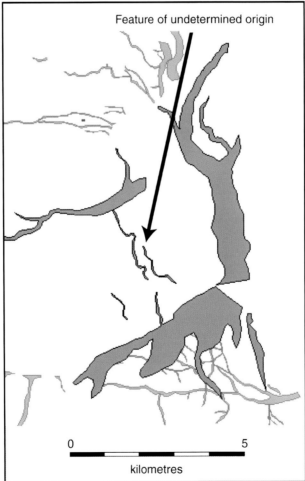

Figure 7.14 The location of one of the small structures that resembles a palaeochannel. The origin of these features remains to be determined

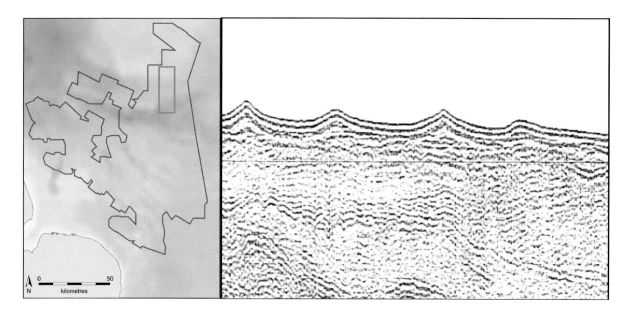

Figure 7.15 Modern Sandwaves directly overlying the Holocene landscape

However, this area would have provided a diverse and productive environment for the occupants of the landscape. The OSP during this period would have hosted a variety of intertidal and estuarine environments within which food would have been relatively abundant.

Two large prominent ridges can be observed in the southern central section of the quadrant (near C in Figure 7.11). These elongate ridge features, have already been discussed in an earlier paper in this volume and were interpreted as moribund sand banks that formed in as estuarine environment during the early Holocene marine transgression (see Briggs et al. this volume; Shennan 2000). From this it is inferred that the tidal range of the OSP was probably macro-tidal and the tidal currents conveyed in the estuary were relatively strong.

7.3.2 Other Features

7.3.2.1 Solid Geology

A series of tunnel valleys can be clearly be observed underlying the Holocene landscape and crossing the OSP within this quadrant (Figure 7.13). These features do not appear to have any topographic expression, but are visible due to the absence of the Later Pleistocene deposits in this depression. These are directly related to features observed on BGS mapping ascribed to the Middle Pleistocene Swarte Bank Formation (Laramine 1989). Due to time and focus constraints only those features that were significant to the interpretation of the Holocene landscape were digitised.

There are, however, a number of small features that resemble a series of palaeochannels. These are located within the Outer Silver Pit, incised into Lower to Middle Pleistocene deposits, and are covered by recent sediment (Figure 7.14).

It is likely that these formed small extensions of tunnel valleys located within this area; although it is equally possible that these are remnants of a more recent palaeolandscape. Unfortunately the origin of these features could not be determined utilising the available data, and the mode of formation and age of these structures remains undetermined.

7.3.2.2 Recent Geological Features

No features of recent geological origin were observed in the seismic data within this region, however 2D BGS seismic lines do show that minor sand structures are located in this area (Figure 7.15).

7.4 South East Quadrant

7.4.1 Description

Few major topographic features were identified within the area (Figure 7.16) and the majority of the surface area within the quadrant was determined to be extremely flat (Figure 7.17). Although the significant data striping present in the southeast of this quadrant hindered interpretation (Figure 7.18), data quality is reasonable elsewhere, and little significant landscape variation is discernable. The southern margin of the OSP is visible in the northern most part of the quadrant (Figure 7.16 B), along with associated intertidal features. A significant depression, which retains a bathymetric expression today, can be observed to the north. However, the majority of the fluvial features within this region run across a large and relatively flat plain. The seismic signal in this area generates a "mottled" appearance, the origin of which is uncertain (Figure 7.18). A few minor fluvial channels can be observed within this mottled zone (Figure 7.16 A). Although detailed interpretation within this area is problematic the area, although undistinguished in topographic terms, may contain significant archaeological potential. The BGS mapping for this region, for example, records an extensive coverage of the archaeologically important early Holocene Elbow Formation over the plain.

Where fluvial features can be defined within this quadrant they can be divided into two groups. The first group is located to the northwest of the quadrant and flow from the southwest to the northeast (Figure 7.16 C). These features run directly across Holocene floodplains but have little observed sinuosity. The project was unable to resolve the channels located within these floodplains. However, given the topography, it is likely that the channels have a higher degree of sinuosity than is visible within the data. Adjacent to the coast, a clearly defined and developed estuary can be seen at the termination of the fluvial channels (near Figure 7.16 B). The character of these estuaries suggests that their formation has been controlled by a series of inundations during a period of rising sea level.

The second group of features is located in the southeast of the quadrant and trend southwest - southeast. They may flow towards a location, suggested by Coles (1998), to contain a deep-water channel (Figure 7.16 A, D). Unfortunately, the noise within this region is enough to hinder the resolution of these features. In one area, near that studied by Praeg (1997), a clear and well-developed sinuous channel system can be discerned. This suggests that if the issue of noise could be overcome, possibly following access to newer surveys, the route of these channels could be resolved.

The most significant landscape feature within the quadrant is the large depression that forms Markham's Hole (Figure 7.16 E and Figure 7.19). Located in the northeast of the quadrant, this large, partially in-filled valley retains a bathymetric expression. The seismic data actually reveals that this feature is much deeper than the bathymetry suggests (Figure 7.19).

BGS cross-sections for the area, along with 2D seismic data made available to the project, suggests the existence of significant deposits within this structure. These deposits can be directly related to the Late Pleistocene Botney Cut Formation and are directly overlain by sediments of recent origin. A channel system attached to the end of this tunnel valley, which is incised into the Late Pleistocene Boulders Bank Formation, can be observed to terminate at the Early Holocene coastline. These relationships suggest that this feature dates to the Late Pleistocene or Early Holocene, and strongly suggests that this feature may have been a drainage channel for this depression. The depression may have contained a lake during the early Holocene. This interpretation is supported by the BGS mapping for the area that records the presence of Late Weichselian to Holocene glacio-lacustrine deposits throughout this depression (Brown 1986).

Further to the north, in an area bound by Markham's Hole, the Botney Cut and the OSP, are a series of channels traversing a low-lying area (Figure 7.16 F). The area itself is clearly defined but does not possess any specific characteristic other than the presence of these channels. This might suggest a delta system (Figure 7.20) however there are no structures visible in the seismic data to support this interpretation. Its relatively flat prospect and position, adjacent to one of the OSP marine inlets, suggests that the area might have been a salt marsh for part of its history at least. The channels appear to show two distinct formation phases. A primary, slightly larger, channel may be older, whilst a series of smaller channels may form a network over and around this structure at a slightly later date. Once again, this sequence may be the consequence of rising sea levels.

The OSP coastline is still well pronounced in this area, although the basin appears to widen as it nears the area of the postulated salt marsh. The coastline is characterised by a strong seismic response although tidal scour marks, which are clearly visible elsewhere, are not clearly defined in this area. This might reflect a difference in tidal flow at this point. A deeply incised inlet can be seen in the intertidal area, adjacent to the postulated salt marsh. The origin of this feature is a partially filled glacial tunnel valley, which was inundated during the marine transgression at around 9,500 BP.

7.4.2 Other Features

7.4.2.1 Recent Geological Features

In the west of the study area are a series of very small linear features. Their position, high in the seismic column and their structure revealed that these features are reflections of large sand waves on the seabed surface and are of recent origin.

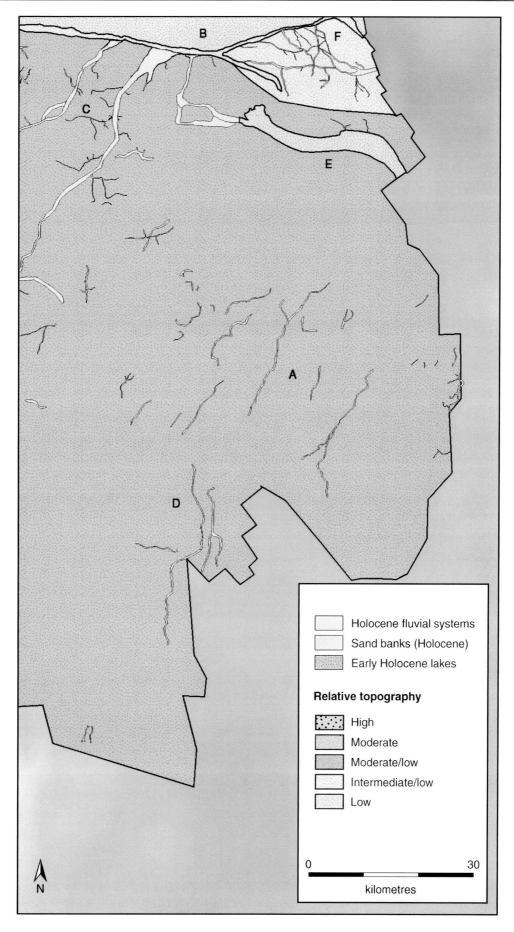

Figure 7.16 The Holocene landscape and features within the southeastern quadrant

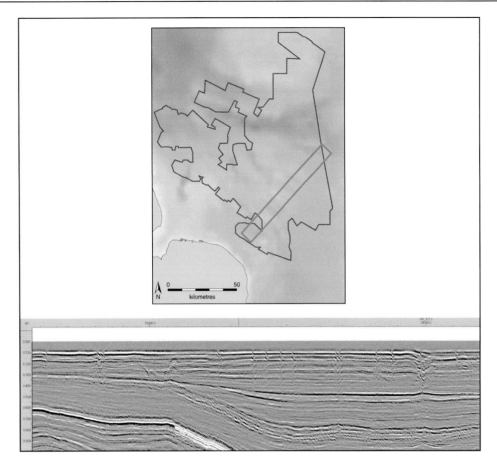

Figure 7.17 Seismic line across the southeastern quadrant. Little landscape variation can be discerned in this area

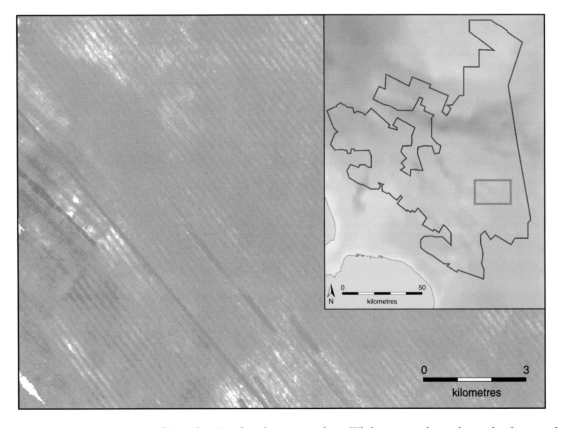

Figure 7.18 A representative image of "mottling" within the seismic data. Whilst minor channels can be discerned, the clarity of the interpretation is hindered.

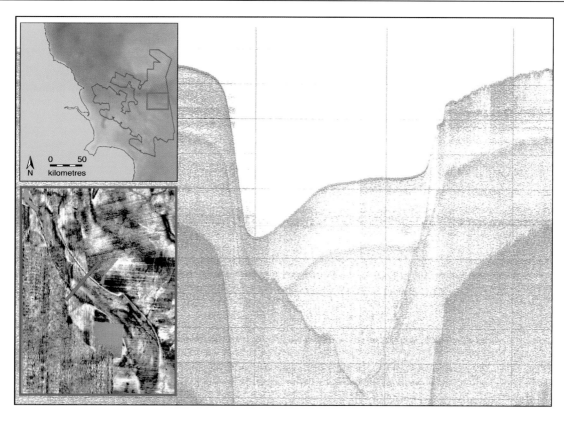

Figure 7.19 Cross section (BGS line 80-01-05) over Markham's Hole. This clearly shows the deep incision of this feature (location marked by red line on inset). The inset timeslice of this feature shows the location of the outflow channel (marked orange)

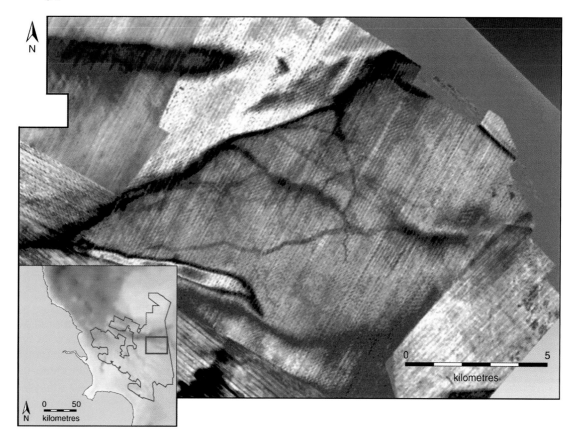

Figure 7.20 Seismic timeslice image of the palaeochannels within an area interpreted as a salt marsh

7.5 South Western Quadrant

7.5.1 Description

The landscape which has been observed within the seismic data for this quadrant reflects the variable influence of the underlying geology upon the observed siesmic signal. Areas in which the geological influence is weak are generally associated with the presence of deep Late Pleistocene deposits. However, the overall character of this landscape is as a relatively gentle plain sloping from the modern coastline onto a lower plain in the north and west of the quadrant (Figure 7.21).

Beginning in the east of the quadrant, the results suggest a relatively flat lying terrain, with a number of large, interspersed depressions. These depressions represent unfilled tunnel valleys that may have contained lakes due to their pre-existing low lying topography and basin-like nature. This is supported in part by the presence of lacustrine deposits relating to the Late Plesitocene and Early Holocene within sediments recovered from some of these depressions (Cook 1991). However later erosional events makes the calculation of the full lacustrine extent problematic (Cook 1991). The smaller of these basins probably corresponds to the Well Hole (Figure 7.21 A). However, the shape of this feature only partially corresponds to the feature observable within the 3D seismic data. A cross section through this feature was provided by an additional high-resolution 2D digital line (93-01-81) provided by the BGS (Figure 7.22). This 2D line reveals a partly eroded deposit, which directly overlies the erosion surface but is beneath modern sediments. Similar results are recorded on the BGS maps for the area, which describe the deposits within this feature as pertaining to the Botney Cut Formation. This deposit dates from the latest Pleistocene/earliest Holocene and is associated with a glacio-lacustrine origin. Given that no obvious outflow is observed, this tunnel valley presumably formed a lake within the surrounding flat lying landscape.

The other major depression within this plain corresponds to the Sole Pit (Figure 7.21 B). The available high-resolution 2D seismic data reveals that the base of this pit has also suffered significant erosion, removing any deposits of Late Pleistocene/Early Holocene age. Minor deposits of this material do occur on the flanks of this feature. It is suggested that this feature would have formed a similar lacustrine feature to 7.21 A. Holocene erosion within this feature ensures that confirmation of such an interpretation is likely to be problematic.

Further to the west of the quadrant, significant landscape features reflect the influence of underlying solid geology (Figure 7.21 C). This is indicated by the clarity of solid geological structures within the seismic data and their influence upon the early Holocene landscape. The seismic data indicates a slight, but significant, depression located in the southwestern corner of the quadrant. This is bounded on either side by low but significant scarp slopes.

Geological beds pushed to near-surface positions by the crests of deep underlying salt structures form these slopes. The result of this process would have been a landscape comprising a broad, low valley bounded on either side by gentle slopes. Given that the majority of fluvial features observed within the area appear to drain into this valley (Figure 7.21 C, D), it is suggested that this area formed a low lying wetland plain, occupied by fluvial systems. Within the valley, one large upstanding feature can be observed (Figure 7.21 E). This indicates the presence of an underlying salt dome that has gently raised the geological beds within this area. This would have formed a low hill within a relatively flat area. Such a hill might represent an important locale for hunter-gatherers. Aside from potential for settlement, game could have been observed from this vantage point as it migrated up the valley (e.g. Fischer 2004, 34).

Slightly to the north is a salt dome (right of Figure 7.21 D) where the break of slope coincides, again, with underlying geological formations. A rise is located within the north of this feature and relates to the peak of the underlying salt dome. This feature is considered as indicative of a relatively raised area in comparison to the valley. This would have formed a distinctive topographic high within the contemporary landscape.

The large depression associated with the Inner Silver Pit is located to the southwest of the hill mentioned above (Figure 7.21 F). The seismic data is, unfortunately, heavily striped in this area. Despite this, the broad outline of the feature can be determined and available geological mapping for the area suggests that the depression is heavily incised and that all sediment has been removed from within the bulk of the feature. Whilst BGS mapping indicates an absence of Botney Cut Formation deposits in the area it is possible that patches of sediments may remain within the Inner Silver Pit. Whilst it is impossible to determine if this depression formed a lake or wetland during the early Holocene, the feature is so pronounced today that it is extremely unlikely that it had no topographic expression during the early Holocene. On that basis, it seems reasonable to suggest that the feature would have contained a lacustrine environment of some sort. A series of palaeochannels can be seen to emerge from this feature (between F and C on Figure 7.21). Unfortunately, noise within the data renders it impossible to determine if these features are related to the Holocene or earlier landscapes.

The southwestern quadrant is associated with three principal groups of fluvial systems. The first, located in the northwestern area (Figure 7.21 G), may be equated to those described in the northwest quadrant. They comprise a series of small, truncated channels that appear to be generally poorly preserved. Although some larger sections provide a clearer response their fragmentary nature is almost certainly a product of erosion. The second group includes those features that are a continuation of channel systems observed in the southeastern quadrant (Figure 7.21 H). These also drain towards the northeast, and a few abandoned meander bends are associated with these channels. They appear large and well developed and a cross section through one indicates a substantial sedimentary profile (Figure 7.23). Given the size and significance of these riverine features, they may well have provided an important route through the landscape, and to and from the coast (Barton and Roberts 2004, 352).

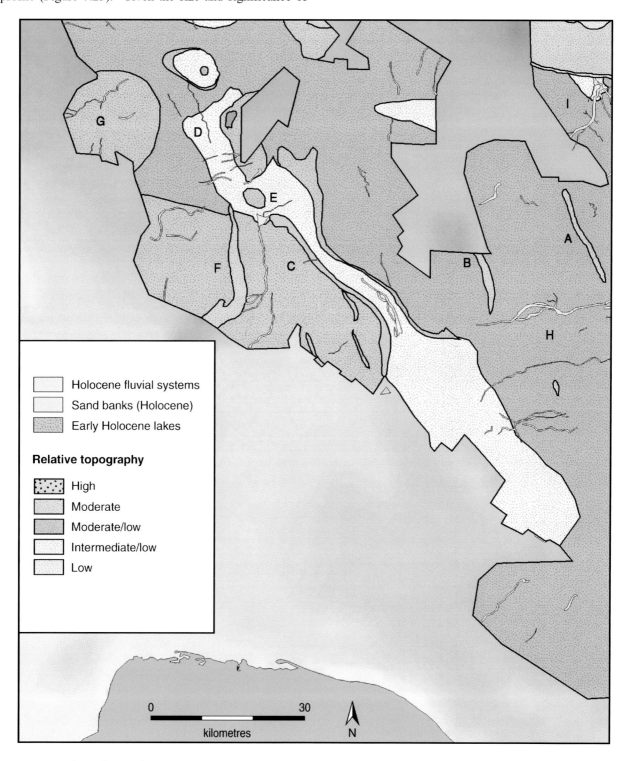

Figure 7.21 The Holocene landscape and features within the southwestern quadrant

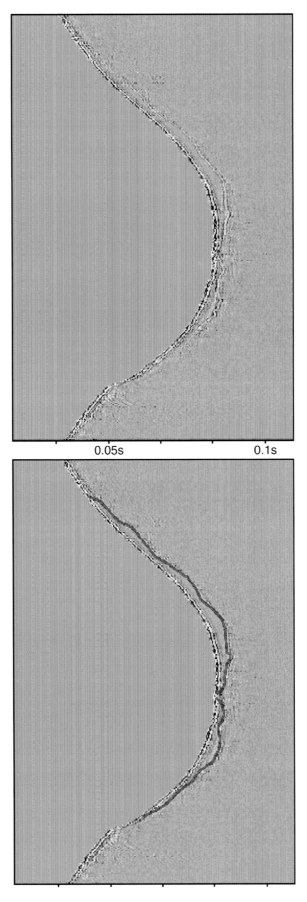

Figure 7.22 Cross section through Well Hole (BGS Line 93-01-81. Note the preservation of the Botney Cut Formation (base marked red on interpreted section)

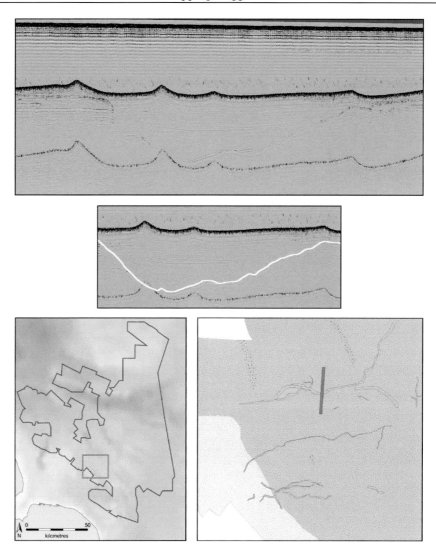

Figure 7.23 Cross section through the large channels in the southwestern quadrant. (BGS Line 93-01-74A)

The third group of features is located in the far northeastern corner of the quadrant, adjacent to the coastline (Figure 7.21 I). These features are similar to those observed in the northeastern quadrant and display dendritic tributaries with well-developed floodplains.

This group displays evidence for change in sea level indicated by the abandonment of part of the channel and the overlying formation of an associated estuary or tidal flat. The estuary has a clear bathymetric expression within the BGS Digbath250 dataset, which suggests that any archaeological deposits may be near the surface. This area could well represent a location with significant potential for future palaeoenvironmental sampling.

7.5.2 Other Features

7.5.2.1 Fluvio-Glacial features

A series of channel structures were observed to lie beneath Holocene features (Figure 7.23). In the west of the quadrant, and near the location of the later depression at Figure 7.21 B, a large fluvio-glacial outwash plain may also be observed beneath the Holocene land surface (Figure 7.24). Although this has little observed impact upon the overlying Holocene landscape, the feature may have archaeological significance as it may contain deposits relating to the earlier Palaeolithic occupation of this landscape.

7.5.2.2 Recent Geological Features

Four main sets of sand ridges were observed in the west of the study area. These sand ridges cluster as a series of parallel lines and are reflected in the upper sections of the seismic column. These are interpreted as large seabed, sand waves of recent origin.

7.6 Conclusions

The mapping and recording of such an extensive archaeological landscape has been a daunting task. The primary statistics relating to the extent of features recorded during the project is provided in Table 7.1.

Table 7.1 Basic quantitative data relating to identified landscape features

Coastline Length Observed	691 km
Marine Area Observed	1791 km²
Lakes/Wetlands Observed	24
Salt Marsh Area	309 km²
Intertidal Zone area observed	293 km²
Major Estuaries Observed	10
Total Fluvial Stream Length	1612 km
Fluvial Related Features Observed	305
Number of Stream Segments	719
Total Area covered by Fluvial Features	526 km
Mean Strahler Order	1.52
Mean Shrever Order	3.64
Average Angle of Stream Join	68 degrees

In considering the results of the NSPP mapping exercise it should be stressed that the interpretation provided here is historic in several senses. The 3D seismic data was acquired over a period of 20 years at least. The PGS Mega survey does not, therefore, represent a snap shot of any single period of time and the information presented here may not represent the full effect of modern changes to the preserved elements of the landscape. However, the effects of burial and erosion across such a vast landscape are likely to be relatively small given the time period under consideration. As such, the general thrust of this chapter remains valid and the data can, with some caution, be used in support of larger archaeological synthesis and for management purposes. However, it is clear that the results of this mapping will require future ground truthing. Coring is required to provide samples for palaeoenvironmental study and dating, whilst higher resolution survey might usefully be considered to resolve specific issues or enhance interpretation.

In conclusion, it seems reasonable to stress that the quantitative data and associated mapping presented here represents a quantum shift in respect of our knowledge of the Holocene landscapes of the North Sea. The study has revealed a hunter gatherer landscape that is, currently, without parallel in Europe and, moreover, may prove to be an optimum area of settlement during the Early Holocene. Interpretation of this landscape in terms of habitation potential and sediment survival will undoubtedly affect our understanding of the regional archaeology of all the countries bounding the survey area. Whilst consideration of the wider archaeological significance of the data will be presented in the final paper of this volume, the results presented here represent a major achievement and reflect the combined efforts of all involved in the project.

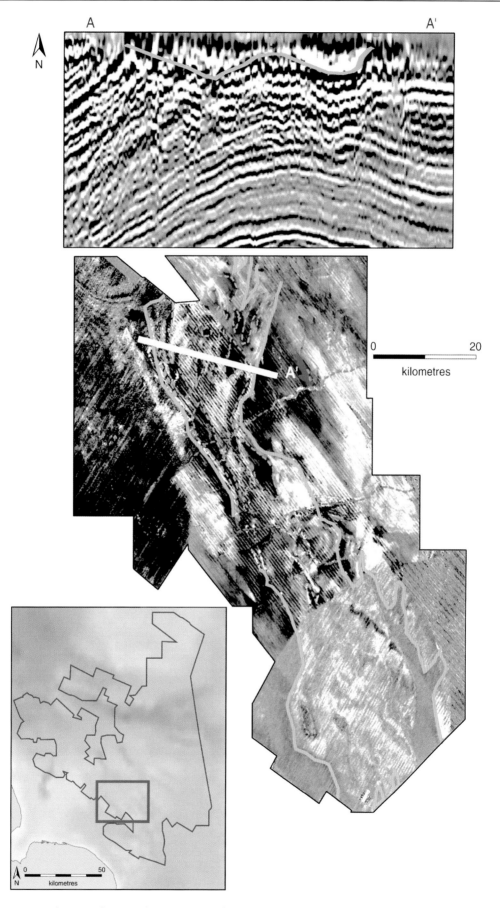

Figure 7.24 An image showing the complex structure of a glacial outwash plain. A cross section through this feature (Line A-A') shows that this feature is located deep beneath the seabed (marked on in green)

8 The Potential of the Organic Archive for Environmental Reconstruction: An Assessment of Selected Borehole Sediments from the Southern North Sea.

David N. Smith, Tom C. B. Hill, Ben R. Gearey, Simon Fitch,
Simon Holford, Andy J. Howard and Christina Jolliffe

8.1 Introduction

Prior to the inundation of the region, during the eustatic sea level rise of the early Holocene, the landscape of the North Sea basin would have presented early human settlers with a range of ecosystems, resources for food and shelter, as well as barriers that restricted their movement. Therefore, the application of appropriate environmental archaeological methodologies, that help to elucidate signals of both natural and anthropogenic landscape change, within a well constrained chronostratigraphic framework, will be integral to the development of any archaeological research framework for the North Sea (Bell and Walker 2005).

The identification of peat deposits in the present intertidal zone fringing the North Sea Basin (e.g. Horton et al. 1999) and further offshore (e.g. Shennan et al. 2000) demonstrates the potential for organic preservation and associated palaeoenvironmental reconstruction using a range of proxy indicators. The analysis and interpretation of 3D siesmic survey data as part of this project has identified a range of natural sedimentary traps capable of containing further environmental remains, such as palaeochannels and floodplain wetlands; however, these need to be 'ground truthed' in order to prove their environmental potential. Such an approach will allow previous interpretations of landscape evolution and archaeological potential of the region to be assessed (e.g. Coles 1998; Flemming 2004; Shennan et al. 2000; Shennan and Horton 2002) as well as aid the development of new models.

Logistically, obtaining environmental samples from beneath the southern North Sea presents many obstacales. For example, delimiting the position of suitable sediment traps and the significant expense of commissioning marine sampling are two of the most obvious issues. However, preliminary enquiries indicated that the British Geological Survey (BGS) has an archive of Late Pleistocene-Holocene material, which they have recovered from shallow boreholes and vibrocores. Therefore, the aim of this part of the project was to assess whether any of the existing sedimentary material stored at BGS offices (National Geoscience Data Centre, Keyworth; Murchison House, Edinburgh) could be used to provide meaningful enviromental evidence from natural features identified in the 3D siesmic survey.

8.2 Potential and Rationale

In the Holocene, river valleys and associated floodplain wetlands of the scale recorded in the 3D seismic survey data are widely recognised as key locations for the preservation of environmental and biological archives (Brown 1997; Howard and Macklin 1999). Classic 'landward' examples of this are recorded by a range of studies from the Great Ouse (Dawson 2000), the Severn (Brown 1983), the Thames (Sidell et al. 2000) and the Trent (Knight and Howard 2005). These sedimentary archives are usually associated with abandoned (in-filled) river channels, wetlands and associated sand and gravel splays, with some spanning considerable periods of time (Brayshay and Dinnin 1999; Smith et al. 2006). However, despite relatively abundant evidence for preservation, the majority of landward archives are associated with middle to late Holocene palaeoenvironments (e.g. Greenwood and Smith 2005; Howard 2005) and reworking of the floodplain may be a key factor in the under-representation of earlier Holocene organic records. In terms of the southern North Sea, it is probable that the channel deposits identified during the analysis and interpretation of the 3D seismic survey data also contain valuable archives of this type, and furthermore, have the potential to fill this missing gap within the early Holocene.

Within the perimarine zone, the potential of intertidal sedimentary archives for environmental reconstruction, including elucidating the nature of sea level change has been clearly demonstrated by recent studies (e.g. Bell et al. 2000); therefore, the potential of now submerged intertidal areas is no less. There is also clear evidence that many early and middle Holocene 'peats' from modern intertidal areas have suffered erosion during periods of later Holocene sea level change (Bell et al. 2000), which has clear implications for the quality of the archive. Furthermore, along areas of the northern British coastline there is growing evidence that extreme events such as the 'tsunami' associated with the Storegga landslide may have resulted in the erosion of environmental archives (Boomer et al. 2007). Though the effects of the Storegga landslide on the contemporary coastal margins of the central North Sea are not clear, it is unlikely that late Holocene and modern erosion will have affected the buried landscapes recorded in the seismic survey. Effectively, this means that the central North Sea may actually have the potential to contain some the best-preserved Early Holocene palaeoenvironmental records for Northern Europe.

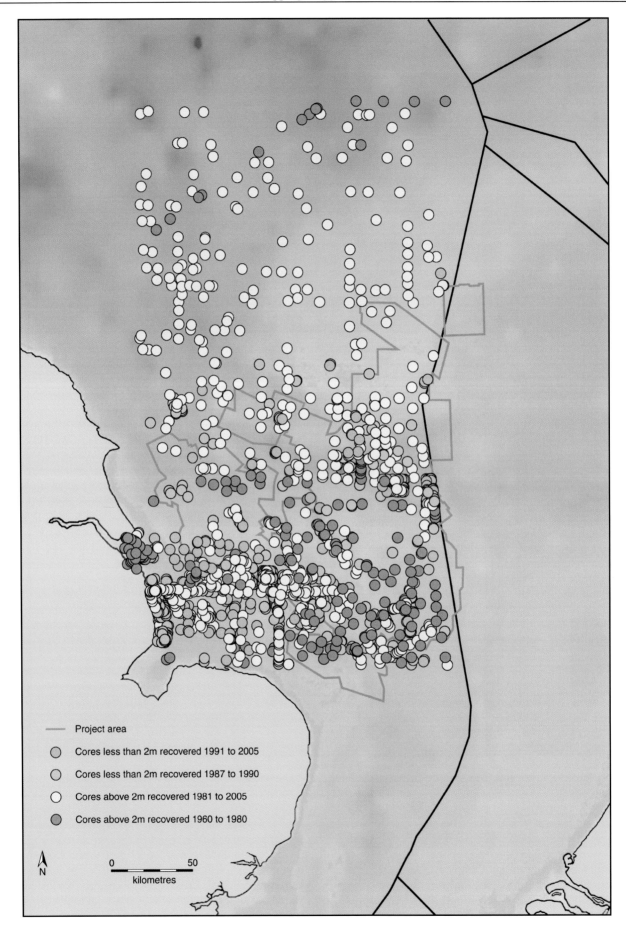

Figure 8.1 Location of the boreholes in the area included in the 3D seismic survey

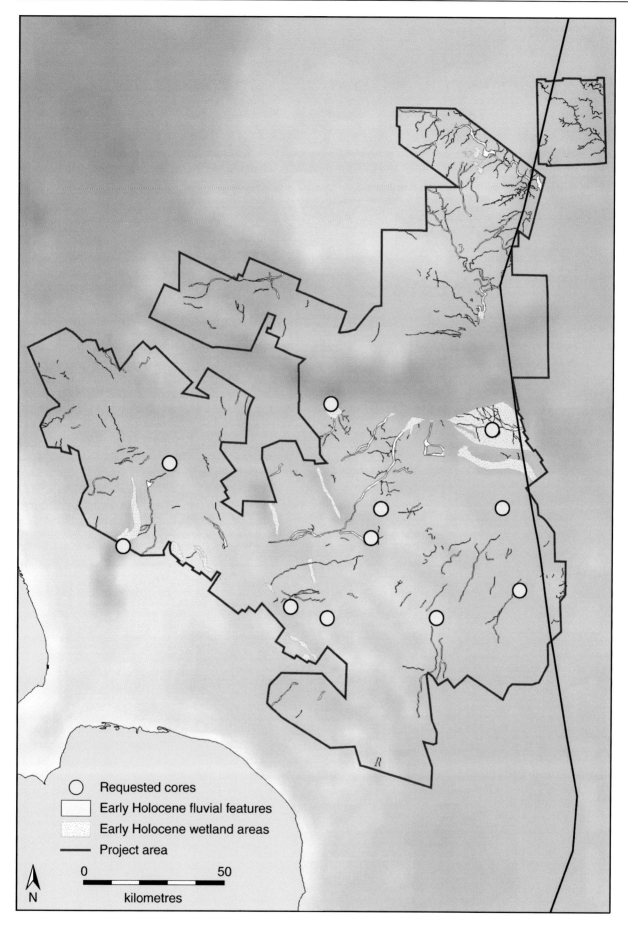

Figure 8.2 Location of the boreholes requested from the area and primary features identified during mapping

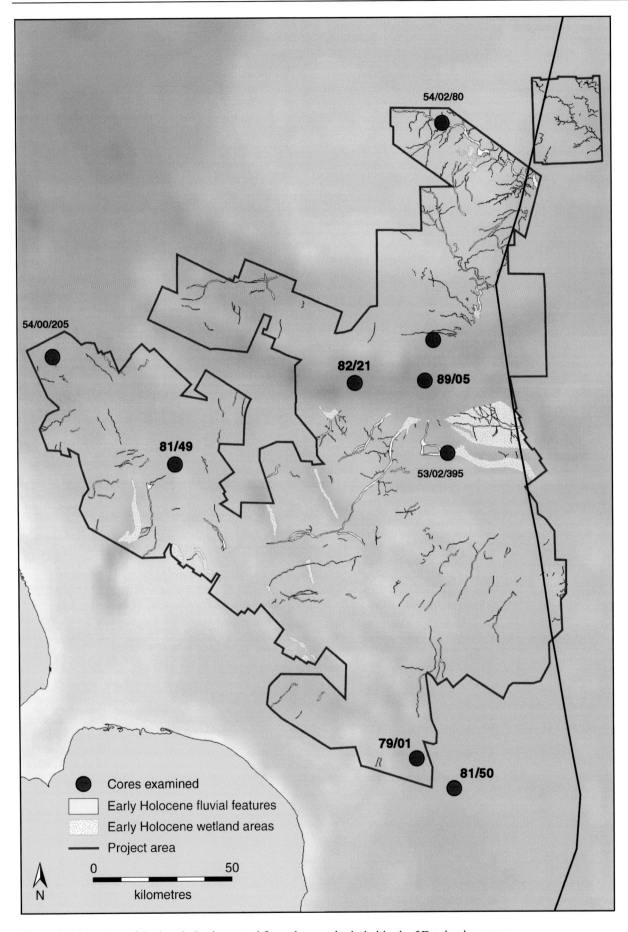

Figure 8.3 Location of the boreholes inspected from the area included in the 3D seismic survey

8.3 Core Selection

Two groups of site-investigation material are held by the BGS in low-temperature storage facilities. The first comprises a series of shallow borehole cores drilled from the 1960s onwards as part of a program of systematic mapping of the offshore geology of the United Kingdom Continental Shelf (UKCS). In total, twenty-nine shallow boreholes were drilled within the UKCS quadrants that cover the area of the 3D seismic survey (Figure 8.1). In addition, vibrocore samples were acquired over this extended time period to aid in the drafting of 1:250,000 seabed sediment maps.

Initially, a sample of nineteen shallow cores and vibrocores were requested from the BGS (Figure 8.2). The selection of these was based on two factors: (1) their proximity to landform features (i.e. sedimentary traps) that had been identified during analysis and interpretation of the 3D seismic data; (2) indications for the presence of 'peat' in these borehole records.

0f the (nineteen) cores requested, six were not available and four existed only as paper archives. As a result, only four shallow boreholes (79/01, 81/49, 82/21, 89/05) and four vibrocores (53/02/395, 54/00/205, 54/02/80 and 54/02/215) from the area under study were available for examination (Figure 8.3). Given the low numbers of cores available, it was therefore decided to also request shallow core 81/50, drilled to the south of the area of the 3D seismic survey (because the field descriptions suggested that it had organic inclusions). Between the 21st and 24th of March 2006, one of the authors (Simon Holford) recorded the nine selected cores at Edinburgh, describing their sedimentary characteristics, condition and potential for further environmental analyses. Previous research had identified the 'Elbow Formation'; a unit of fine-grained muddy sands and interbedded clay of early Holocene date, which is mappable in the eastern and southern part of the project survey area (Cameron et al. 1992; Ward et al. 2006). the key to further work, therefore, was to identify the presence of organic (peat) sediments that might be of Holocene age.

In terms of quality of storage and general degree of preservation, the core samples were found to be in a good condition given their age (the shallow boreholes are named after the year and order in which they were drilled i.e. 79/01 was the first borehole drilled in 1979). However, some surfaces of the cores were covered in mould (sections 89/05 and 54/02/215) and all were largely desiccated. Both of these conditions inhibit the preservation of organic material and mould growth certainly limits any potential to date materials using radiocarbon techniques. Abundant shell fragments of a mollusc (possibly *Spisula subtruncata*) were recovered in many of the cores. However, these fragments were considered unsuitable for detailed further study, such as stable isotope analysis or amino acid racemization dating techniques. A full description of the cores examined and their condition is given in the Appendix.

8.4 Palaeoenvironmental Assessment

8.4.1 Sampling

Sampling of the remaining five borehole cores occurred on the 15th June 2006. After discussions with curatorial staff at the BGS, it was agreed that sampling quotients should be limited to a maximum of four samples per 1 metre section of core and the size taken should be limited to 50g of material. In total, fifty samples were obtained from three vibrocores (six samples from 53/02/395, nineteen samples from 54/02/215 and twenty-three from 54/02/80) and a single sample and a wood fragment from one shallow borehole (81/50). The borehole details, sample numbers and sample depths are summarised in Table 8.1.

8.4.2 Visual Assessment

Samples from vibrocore 53/02/395 (samples NSP01-06) were predominantly medium grained, light yellow-brown sands with a relative abundance of disarticulated shell fragments. Samples from vibrocore 54/02/215 (samples NSP07-25) were fine to medium grained, light yellow-brown sands, which appear to become slightly darker yellow-brown with depth, possibly in response to an increase in humic organic content. There also appeared to be a gradual reduction in abundance of disarticulated shell fragments with depth. Samples from vibrocore 54/02/80 were fine-grained, light yellow-brown sands with occasional disarticulated sand fragments (although in lower abundance than found in the previous two vibrocores).

8.4.3 Assessment of macrofossil (insect and plant) inclusions

Initially, it was intended to use the assessment procedures for insect and plant remains outlined by Kenward *et al.* (1986) and to assess the degree of preservation following Kenward and Hall (2006). All forty-nine samples of material from the boreholes were rapidly scanned under a low-power binocular microscope and no organic macrofossils were observed. The only exception to this was a small fragment of poorly preserved wood from borehole 81/50 (NSP49). Given that the wood came from unconsolidated sands, its context was poorly constrained and the pollen assemblage from the sample was poorly preserved it was decided not to date this fragment.

8.4.4 Pollen Assessment

In total, sixteen samples were selected for the assessment of plant microfossils (i.e. pollen grains, moss fragments and plant spores). This selection was based on the conclusion that the sediment samples from the basal parts of the boreholes and vibrocores afforded the greatest environmental potential, especially since the upper parts of the cores might be reworked. In addition, the slightly darker yellow-brown nature of some of the deeper sediments was also considered to be an indicator of the potential presence of organic remains, which in turn, increases the likelihood of plant macrofossil preservation.

Table 8.1 Summary of sedimentary samples and environmental assessment undertaken from the Southern North Sea vibrocores and boreholes

Borehole/core name	Sample number	Depth	Pollen Analysis	Beetle Analysis	Plant macrofossil Analysis
53/02/395	NSP01	6.68 m	*	*	*
53/02/395	NSP02	7.13 m	*	*	*
53/02/395	NSP03	8.24 m	*	*	*
53/02/395	NSP04	9.13 m	*	*	*
53/02/395	NSP05	10.23 m	*	*	*
53/02/395	NSP06	11.25 m	*	*	*
54/02/215	NSP07	0.2 m		*	*
54/02/215	NSP08	0.4 m		*	*
54/02/215	NSP09	0.6 m		*	*
54/02/215	NSP10	1.2 m		*	*
54/02/215	NSP11	1.5 m		*	*
54/02/215	NSP12	1.8 m		*	*
54/02/215	NSP13	2.2 m		*	*
54/02/215	NSP14	2.5 m		*	*
54/02/215	NSP15	2.8 m		*	*
54/02/215	NSP16	3.2 m		*	*
54/02/215	NSP17	3.5 m		*	*
54/02/215	NSP18	3.8 m		*	*
54/02/215	NSP19	4.2 m	*	*	*
54/02/215	NSP20	4.4 m		*	*
54/02/215	NSP21	4.6 m	*	*	*
54/02/215	NSP22	4.8 m		*	*
54/02/215	NSP23	5.2 m	*	*	*
54/02/215	NSP24	5.4 m		*	*
54/02/215	NSP25	5.6 m	*	*	*
54/02/80	NSP26	0.4 m		*	*
54/02/80	NSP27	0.6 m		*	*
54/02/80	NSP28	0.8 m		*	*
54/02/80	NSP29	1.2 m		*	*
54/02/80	NSP30	1.4 m		*	*
54/02/80	NSP31	1.6 m		*	*
54/02/80	NSP32	1.8 m		*	*
54/02/80	NSP33	2.2 m		*	*
54/02/80	NSP34	2.4 m		*	*
54/02/80	NSP35	2.6 m		*	*
54/02/80	NSP36	2.8 m		*	*
54/02/80	NSP37	3.2 m		*	*
54/02/80	NSP38	3.4 m		*	*
54/02/80	NSP39	3.6 m		*	*
54/02/80	NSP40	3.8 m	*	*	*
54/02/80	NSP41	4.2 m		*	*
54/02/80	NSP42	4.4 m	*	*	*
54/02/80	NSP43	4.6 m		*	*
54/02/80	NSP44	4.8 m	*	*	*
54/02/80	NSP45	5.2 m		*	*
54/02/80	NSP46	5.4 m	*	*	*
54/02/80	NSP47	5.6 m		*	*
54/02/80	NSP48	5.8 m	*	*	*
81/50	NSP50	11.9 m	*	*	*

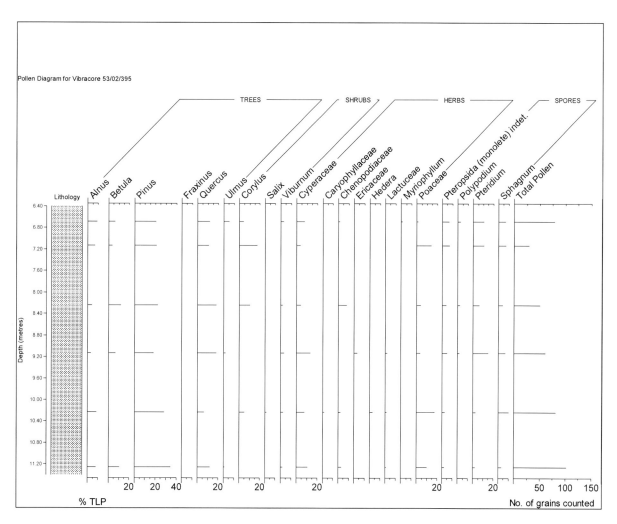

Figure 8.4 Pollen diagram for Vibrocore 53/02/395

Preparation for pollen assessment followed the standard techniques including KOH digestion and acetylation (Moore *et al.* 1991). Due to the high inorganic content of all samples, a very small amount of residue remained once pollen preparation had been completed. Each sample was mounted on a 24x40mm coverslip and glass slide for counting. The pollen assessments were carried out on a Meiji MX5000 microscope at x400 magnification.

To ensure a statistically valid environmental assessment is achieved, pollen analysis normally requires a minimum of 125 total land pollen grains (TLP) excluding aquatics and spores to be counted for each sample. However, pollen occurrence was very low in all samples, resulting in the minimum 125 TLP not being achieved in any single sample. A complete microscope slide was therefore traversed for each sample in order to assess species abundance, diversity and conditions of preservation. Pollen nomenclature follows Moore *et al.* (1991).

8.4.5 Results

The results are presented as pollen diagrams produced using the computer program TILIA (Grimm 1991). The diagrams display percentage values rather than concentrations since these were universally low. None of the pollen spectra recovered reaches the recommended assessment count of 125 let alone the 350 recommended for full analysis. As a result there is no attempt here to use this data to reconstruct prevailing vegetation or to reconstruct landscape.

8.4.5.1 Vibrocore 53/02/395 (Figure 8.4)

Six samples (6.68, 7.13, 8.24, 9.13, 10.23 and 11.25m along the core) were assessed. The low counts of pollen recorded preclude detailed interpretation. Pollen occurrence was low although increased with depth. Trees and shrubs are well represented (80%+ TLP), initially dominated by *Pinus* (pine) and *Quercus* (oak) with small amounts of *Alnus* (alder) and *Betula* (birch). *Corylus* (hazel) increases in abundance vertically through the sequence. *Ulmus* (elm) and *Salix* (willow) are recorded at trace values. Herbs are recorded in the form of Poaceae (wild grasses), Cyperaceae (sedges) and Chenopodiaceae (fat hen), whilst *Pteridium* (bracken) and *Sphagnum* (bog moss) spores are also present, the former increasing down the profile after 9.13m.

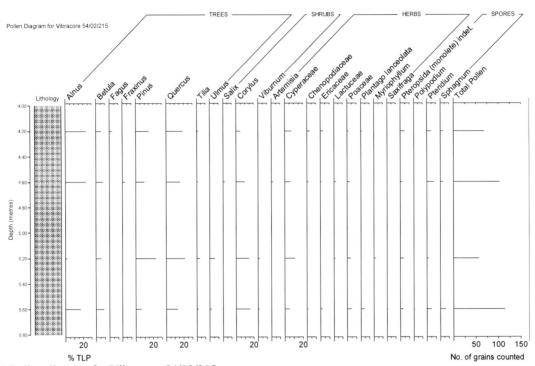
Figure 8.5 Pollen diagram for Vibrocore 54/02/215

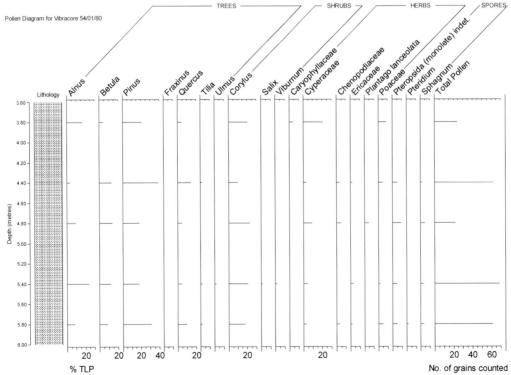
Figure 8.6 Pollen diagram for Vibrocore 34/02/80

8.4.5.2 Vibrocore 54/02/215 (Figure 8.5)

In total, four samples were assessed for pollen from vibrocore 54/02/215. All samples were taken from towards the base of the core profile; 4.20m, 4.60m, 5.20m and 5.60m (Figure 8.5). The impression is of mixed dense woodland with oak, pine and hazel into which alder seems to have subsequently expanded. Tree and shrub pollen (85%) again dominate the spectra, initially with pine, oak, hazel and birch, with lower values for *Tilia* (lime), *Ulmus* (elm) *Fraxinus* (ash), Ericaceae (heather family) and *Salix*. *Alnus* increases above 4.60m with *Fagus* (beech) also appearing in the uppermost sample. Clearance and openings in the woodland are suggested by trace values for Poaceae (grasses) and Cyperaceae (sedges), along with *Plantago lanceolata* (ribwort plantain), Chenopodiaceae, Lactuceae (dandelions etc.) and *Artemisia* (mugwort). Spores in the form of *Pteridium*, *Sphagnum* and Pteropsida (fern) are present across the spectra. Both the scale and causes of these clearances cannot be speculated upon.

8.4.5.3 Vibrocore 54/02/80 (Figure 8.6)

Five samples were assessed for pollen from vibrocore 54/02/80. All samples were taken from towards the base of the core profile; 3.80m, 4.40m, 4.80m, 5.40m, 5.80m (Figure 8.6). Pollen concentrations were again generally low; however despite this, pollen preservation was reasonable for most of the samples. Tree and shrub pollen dominate the spectra (90%+TLP), with *Alnus*, *Betula*, *Pinus*, *Corylus* and *Quercus*. Trace values are recorded for *Tilia*, *Ulmus* and the shrubs *Salix* and *Viburnum*. Few herbaceous taxa are present, although low values for Poaceae and Cyperaceae increase slightly towards the top of the sequence. The only other herbs present are *Plantago lanceolata*, Chenopodiaceae and Caryophyllaceae (pink family). *Sphagnum* and Pteropsida spores are found in low quantities towards the base of the core.

8.4.5.4 Borehole 81/50

At a depth of 11.90m in borehole 81/50, a wooden fragment was observed within the sedimentary profile. This sample was therefore subsequently also assessed for pollen. Pollen abundance was extremely low within the sample, with only eight pollen grains counted. As a result, pollen concentrations were too low to permit any tentative conclusions to be drawn.

8.5 Discussion

The generally low occurrence and apparent concentration of pollen in all the sequences discussed preclude detailed interpretation. All of the spectra recovered commonly contained grains that showed evidence of fragmentation, distortion and occasionally corrosion. Species identification was commonly hindered as a result of the poor pollen preservation. It is possible that this poor preservation represents the degradation of material in storage after sampling. However, it is probably more likely that it represents sediment reworking, with previously deposited pollen grains being re-eroded and subsequently deposited within the sedimentary archive.

The overall vegetation reflected is generally similar for all three sequences, although some differences are observed, such as slightly higher values for grasses and sedges in vibrocore 53/02/395. The dominance of tree and shrub taxa including pine, birch, oak and alder suggests a Holocene timeframe. Without an adequate radiometric dating framework, it is impossible to provide a precise age but the pollen spectra do suggest an early to middle Holocene date. For example, lime is absent from these cores but is relatively common after the mid-Holocene (Huntley and Birks 1983). Elm, which is recorded in some of the samples, is usually present at higher values in the early part of the Holocene (eg. Huntley and Birks 1983). However, given the considerable chance that taphonomic/depositional bias, local edaphic factors or problems of pollen production and representation could be patterning these very small spectra this provisional age estimate should be viewed with caution.

The presence of grasses and sedges does suggest the presence of open habitats within these woodlands. Whilst these openings in the canopy may reflect areas of wetter soil, well known indicators of 'clearance' such as ribwort plantain and dandelions, are encountered in vibrocore 54/02/215 and vibrocore 54/02/80. However, given the constraints of the data set, it is perhaps unwise to make detailed comment regarding the possible role of human communities in the creation or exploitation of any such open areas.

8.6 Conclusions

There is no doubt that deposits associated with the landscape features identified in the southern North Sea have significant potential to provide proxy records of past vegetation, climate and land use history, which can be placed within a secure chronostratigraphic framework. Unfortunately, this environmental assessment suggests that the existing late Pleistocene-Holocene sedimentary archives retained by the BGS have limited value to address these research questions. Despite being well curated, they do not appear to contain the type of material needed for this study. This is not surprising since these cores were taken for distinctly different research purposes to those of this current survey. Equally, they were drilled before detailed palaeogeographic maps of the resolution provided by this project were available and hence do not appear to have been extracted from the most promising palaeoenvironmental localities.

In order to provide more substantial palaeoenvironmental understanding of the environmental archive within the North Sea, further 'targeted' sampling from the seabed is necessary. The locations of any future borehole drilling and sediment sampling programme should be pre-determined by reference to the features identified within the 3D seismic atlas. Promising, major landform features for initial targeted sampling include the 'Shotton River' and associated wetlands and salt marshes. Any further environmental analysis should be undertaken within a well-developed radiometric dating framework and immediate palaeoenvironmental assessments should be undertaken prior to storage in order to prevent sample loss through decomposition and/or desiccation and to stabilise any environmental evidence for future analysis.

Appendix: Sediment and preservation descriptions of the Holocene / Late Pleistocene sections of the cores examined in this survey

Shallow boreholes

79/01

This borehole was located at the southern extreme of the PGS southern North Sea 3D seismic survey. Unfortunately the drilling of this borehole was dogged with technical problems (completion reports, Murchison House) resulting in limited recovery of late Pleistocene-Holocene deposits. The inspected sediments were held in sample jars and comprised mostly brown, shelly sand. It was therefore concluded that material from this borehole had limited potential.

81/49

This borehole is located in the western part of the PGS megasurvey. A BGS completion report for this borehole stated that between 0 and 11.3m it contained deposits of Pleistocene-recent age. Examination of the available core material over this depth range revealed it to be generally very fine-grained, soft sand-mud deposits, dark grey-brown in colour, often interbedded with fine clay laminae. Towards the base of 'Pleistocene-recent' succession the sediments are more consolidated, desiccated and richer in sand, lacking the clay laminae observed at shallower levels. The entire 'Pleistocene-recent succession' is lacking in shell fragments or organic matter. At around 11m there is a clear transition to dark grey, indurated mudstones of Jurassic age. In summary, no material of potential palaeoenvironmental use was found to be present in this borehole.

81/50

This borehole is located outside the extent of available seismic coverage, 48 km east of Lowestoft. However, a BGS report on both onshore and offshore boreholes acquired during 1981 stated that this borehole contained a horizon of wood fragments at shallow levels (<15 m) within the Pleistocene-recent succession. Between depths of 0 and 10m, yellow-brown, medium-grained unconsolidated sands, often with abundant shell fragments, characterize the retrieved sediments. In places the core is in quite poor condition probably resulting from the unconsolidated material shifting. In places the core has also clearly been extensively sampled. Inspection of the core at the depth of 11.82 to 12.0m found only a few small wood fragments. The apparent paucity of wood fragments is probably a function of previous sampling efforts.

82/21

This borehole was selected for inspection due to its location within the Outer Silver Pit. Overall, the core samples from this borehole were found to be in excellent condition (with some local exceptions where the core material was comparatively desiccated and fragmented. Between 0 and 6m, the inspected sediments comprised mostly dark olive-grey, fine-grained shelly sands and gravels. Below 6m there is a transition to interbedded fine sands and silty clays, with grey scattered shells. Organic materials were not observed at any levels within the examined section of core, even within the more clay-rich units.

89/05

This borehole is located several km to the west of 82/21, and hence was chosen for examination for similar reasons. The completion log stated that this borehole encountered Holocene sediments (mostly muddy sands) down to depths of *c.* 14 m., below which it penetrated sands of Middle Pleistocene and earlier date. The first few metres of the Holocene succession, containing homogenous, soft muddy sand (but no observable organic matter) was in excellent condition. Between 1.5 and 10.0m however, much of the core material was covered by variable amounts of mould raising the possibility that any organic material present, could have been contaminated. Although the presence of numerous wood fragments within the Middle Pleistocene succession had been noted in the completion log, it was apparent upon inspection that this borehole had been extensively sampled for micropalaeontological analyses at an earlier date leaving few horizons containing organic matter.

Vibrocores

54/02/80

This vibrocore was an obvious candidate for examination given its proximity to the large NW/SE trending early Holocene fluvial system recorded by seismic data (the so called Shotton River channel). This vibrocore recovered 5.8m of sedimentary deposits, most probably all Holocene in age. The sediments are in a good condition and comprise mostly fine-grained well-sorted sands, grey-brown in colour with some bands of shell fragments, and slightly muddy layers that may contain datable organic material.

53/02/395

This core was obtained near the southern margin of the Outer Silver Pit. 5.98m. of core were recovered as grab samples, and 5.90m of continuous core were recovered below this. Lithologically, the sediments are very similar to vibrocore 54/02/80; yellow-grey, fine-grained bioturbated sands with fragmented mud laminae and scattered shell debris. Some dark horizons, possibly containing organic material were identified, but no definitive wood fragments or 'peat' horizons were observed. The core samples were generally in good condition, but some sections were covered with mould.

54/02/215

This core was drilled on the opposite side of the Outer Silver Pit. As well as a grab sample, 6m of continuous core were available for inspection. Interestingly, the accompanying BGS report described the vibrocore as having penetrated "probably into the top of a small palaeovalley in this area". The vibrocore samples are best described as soft, olive-coloured muddy sands, well sorted and fine grained. No organic material was observed. Although the cores were in a largely excellent condition, again some sections were covered with mould.

54/00/205

This vibrocore was drilled in the extreme east of the study area. Whereas most of the vibrocores were soft and damp, the sediments from 54/00/205 were completely dry and desiccated. Overall, the sediments are best described as olive-brown fine-grained sands. The shallower units often contain abundant shell fragments and pebbles, although the quantities of both decrease with depth. Again, no material deemed suitable for palaeoenvironmental analyses was found during the examination of these deposits.

9   The Archaeology of the North Sea Palaeolandscapes

Simon Fitch, Vincent Gaffney and Kenneth Thomson

9.1   Introduction

The map data generated as part of this project represents one of the largest samples of a, potentially, well preserved early Holocene landscape surviving in Europe and it is essential that some consideration of the archaeological context of the mapped remains is presented here. The European cultural period associated with this landscape is the Mesolithic which lasts between c. 10,000 BP and c. 5,500 BP, dependent on geographic position. Tremendous environmental change forms the backdrop to cultural events throughout this period. Sea level rise, associated with climate change, resulted in the loss of more than 30,000 km^2 of habitable landscape across the southern North Sea basin during the Mesolithic alone, and the inundation of this immense area has essentially left us with a 'black hole' in the archaeological record for northwestern Europe as a whole. This situation is made worse by the fact that finds from the region only rarely possess an accurate provenance or context (Koojimans 1971; Verhart 2004).

Whilst the Early Mesolithic (10,000 to 8,500BP) record from the North Sea region is, essentially, a blank, the terrestrial record does provide some insight into what might be expected within the area of the Southern North Sea itself (Jacobi 1973; Wymer 1991). The early stages of the English Mesolithic are best represented in England by a small number of sites including Star Carr, Thatcham, Broxbourne and Horsham (Clark 1972, Healy et al. 1992, Warren et al. 1934, Jacobi 1978). These sites do show some variation in culture indicative of differing social groupings (Reyner 1998). In the past there has been a trend to group British sites with those of the Maglemosian of Denmark, and frequently to see parallels with the "Duvensee" culture (Clark 1975). However, there are difficulties with such comparison and they add little to our understanding of the archaeology of the North Sea region as it stands.

In general terms, all of these early sites demonstrate utilisation of a range of resources, primarily focused upon game animals and plants. The Early Mesolithic in England does not yet record substantive evidence of the use of marine resources. However, given the emerging knowledge of coastal change it is likely that the majority of the areas that might record such economic practices are actually submerged. Evidence from Scandinavia, where substantial areas of early coastline survive, suggests that these resources would not have been ignored (Norqvist 1995). Conventionally, the Early Mesolithic has been seen as period where populations moved between base camps on the coastline to inland camps to forage (Clark 1972, Smith 1992, Fischer et al. 2004). This interpretation suggests that the emergent landscape of the SNS should possess seasonally visited base camps during the Early Mesolithic. However, information from Scandinavian suggests a contrasting lifestyle utilising only resources within a maritime zone (Indrelid 1978:169-70, Nygaard 1990:232), and it is possible that the contemporary occupants of the "Doggerland" coastline might have followed a similar lifestyle. If this comparison were correct it would contrast with conventional models of Mesolithic movement for England at least (Darvill 1995, figure 20; Smith 1992). Indeed, whilst it must be acknowledged that previous models have rarely had access to data from the original coastlines, recent discoveries at sites including Howick in Northumberland (Waddington et al. 2003), suggest that we might expect significantly more complexity and diversity in economic and social practise than previously imagined or currently experienced. Any enhancement of our knowledge derived from the submerged landscape of the North Sea is therefore likely to provide information that will significantly refine our appreciation of the Early Mesolithic within the larger region.

The Later Mesolithic in Britain (8,500 to 5,500BP) has often been interpreted as a time of economic change and increasing divergence from cultural developments in Europe (Jacobi 1973; Wymer 1991). Jacobi (1976) for example, concluded that such discrepancies were related to the submergence of parts of the North Sea and the increased difficulty of maintaining connections between Europe and Britain. Certainly, the effect of the final inundation of the North Sea emergent landscape during the Later Mesolithic would have been significant to the many groups who must have lived on, or adjacent to, the North Sea plain. As the historic landscape was gradually lost to the sea the area would have fragmented into islands. Whilst some of these isolated areas, at least, would have continued to be populated as marine transgression progressed, habitation of this region would have become increasingly tenuous and migration from the region must have occurred (Coles 1999). The consequences for the groups who moved, or for those who lived in the areas into which they migrated, are largely unknown and only rarely considered. Whilst some consideration has been given to the issue of migration to areas including Norway and Scotland, for the earlier period of inundation (Nummedal 1924, Bjerk, 1995; Fuglestvedt 2003, Warren 2005, 37), the significance of population movement during the final periods of flooding has hardly ever received consideration (Coles 1999, 54).

The isolation of Britain that is assumed to have derived from these changes is often stressed in the literature. The absence of formal burial in the British archaeological record, for instance, is notable and suggests a cultural difference. It may be that there were separate customs regarding burial in Britain but it is equally possible that formal burial sites do exist and that these may have been located near the coast, in areas which have now been lost to the sea (Barton and Roberts 2004, Chatterton 2005, 108). However, as Funnell (1995) and Coles (1998) observe, Britain did not become an island until c. 5,500BP and the actual effect of the North Sea as a barrier to cultural contacts must be open to question. The use of major river systems for communication seems uncontroversial (Roberts 1987, Reyneir 1998), and it is not inconceivable that the shallow marine areas of the North Sea could have been traversed and contact with European groups maintained (e.g. Coles 1998, 76). One might even suggest that the potential for communication by boat, via shallows, might have actually enhanced the potential for contact rather than acted as a barrier. Consequently, whilst the overall picture provided by the available evidence for the Late Mesolithic within Britain suggests a mosaic of localised groupings we should be cautious when assuming that this reflects enforced isolation (Morrison 1980).

Another traditional characteristic associated with the transition from the Early to Late Mesolithic, and often assumed to be a consequence of the change in sea level, is the assertion that there is an increasing focus upon coastal resources (Rowley Conwy 1983). This shift has been interpreted as a response to higher population levels or mobility caused by sea level rise (Mithen 1999). However, this period is characterised by an increasing visibility of activity in landscapes that had previously been under-represented in the archaeological record, e.g. estuaries which had lain beyond the contemporary coastal margins. These unexplored areas provided a diverse range of resources that were unlikely to be ignored during any period of human occupation (Allen 1997, Clarke 1978). Once again, we should be cautious about the extent, or significance, of such apparent change (Milner 2004). The current picture may well be the result of increased visibility of coastal resource utilisation rather than substantive economic change. The provision of information that permits adequate comparison or assessment of development during this period is again predicated on the availability of representative data for the period overall. On that basis, the potential for the North Sea to provide critical information for such an assessment seems clear.

The probable significance of the results derived from the NSPP mapping of the Holocene surfaces should be clear from the previous discussion. It is essentially true that our current interpretative position for the Mesolithic of the maritime regions of north western Europe stand largely as a consequence of the lack of information from the North Sea. The ability to identify significant landscape features within the area of the North Sea, or information that can support directed exploration or future data gathering, is therefore a considerable opportunity for research. It would also be a serious challenge for the heritage communities within all the countries that bound the North Sea basin (Maarleveld and Peeters 2004). The nature of this challenge can be assessed initially by considering the potential of the new mapping for assessing the character of the archaeological record and, in particular, the potential survival of palaeoenvironmental data. Following this one can assess how current management options may be changed or adapted to use the new data and, finally, it will be necessary to discuss the potential of the data to plan research strategies that may begin to answer some of the research questions outlined in the previous section.

Clearly the resolution of the data produced by the NSPP does not permit a fine-grained assessment of the archaeology of the area. However, many of the features identified through the analysis have the potential to achieve archaeological significance (Table 9.1). Paramount amongst these features is the OSP. This basin dominates the mapped landscape both in extent but also the manner in which so many other features are linked to or drain into it. It also represents, of course, a major economic resource: whether considered a lake or a marine outlet. Surrounded by nearly 700 kilometres of coastline, or lakeshore, merging with 10 major estuaries and a salt march covering more than 300 square kilometres, the OSP must have acted as a prime economic resource for human groups across a massive area. Waterfowl, fish and other animals must have been abundant in this area, as would reeds or other vegetational resources that hunter gatherer groups might require. In its later incarnation as a marine estuary the OSP also provided a significant point of access to the marine resources missing from much of the English terrestrial archaeological record. Presumably this is an area where we could seek evidence for intense utilisation of marine resources and any differing social and settlement structures that might result from access to such resources. Away from this imposing area of water, the twenty four lakes or wetlands and the 1,600 kilometres of rivers or streams recorded by project staff would also have provided similar opportunities for hunter gatherer groups, whilst also directing paths and tracks through the landscape. These features achieve further significance as volumetric and sedimentary analysis suggests that many of these areas are most likely to provide palaeoenvironmental evidence. It may be that we will never be able to explore settlement associated with these features but proxy evidence for settlement and land use gained through a programme of directed coring planned on the basis of the results presented here is a real possibility.

The character of this area cannot, of course, be represented simply as a sequence of river and lakes. It was a landscape in the fullest sense and we must not sidestep the responsibility of treating it as such. Figure 9.1 provides a general interpretation of the mapped landscape data made available through the project. Here it is important to stress that whilst the character of the area studied was, essentially, a plain it was never a featureless plain. Mesolithic communities of the North Sea would have been sensitive to the economic and social significance of this subtle variability.

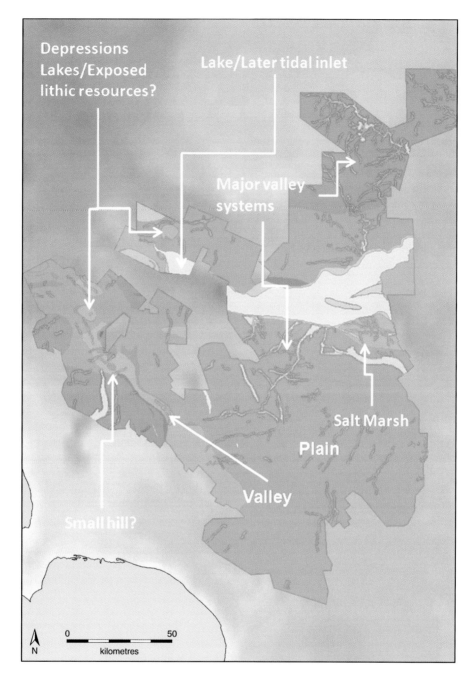

Figure 9.1 Major topographic or economic zones within the study area

Although the current work was only able to provide a broad outline of topography away from major incised features, this remains an achievement in the light of how little we knew previously. We can guess that the large, low valley in the west of the study area would have been attractive for a variety of reasons (figure 9.1). The low hill, tentatively identified within this valley, suggests opportunities for settlement or even for hunting stands. Features associated with salt domes are also particularly interesting in landscape terms. In some cases upswelling domes would have formed low hills but, where there is evidence for graben collapse, the centre of these domes may have contained wetlands or lakes. Such areas are, of course, attractive for human groups and there is at least a chance that faults in these areas might also expose other useful resources including lithic sources.

Contemporary populations, of course, would have understood the great North Sea plain in a much more intense and personal manner. Small groups would have been intimate with features of the landscape that we cannot detect using current technologies. The emotive relationship between individuals and their surrounding landscape can hardly be understood through a study that can only vaguely discern the trend of the land or map its grosser features. It is, however, the best we possess and on that basis we must consider what implications may be drawn in respect of further research.

9.2 Future Research

Despite the limitations outlined above it is incontrovertible that the data presented as part of this project has demonstrated the potential of marine, remote sensed data for the exploration of the inundated Holocene land surfaces of the North Sea. In comparison to the situation described by Flemming, a mere 3 years ago, the North Sea is no longer *terra incognita* (Flemming 2004). It is, of course, acknowledged that the current product still represents a limited interpretation and could be substantially refined by the integration of further data sets, including high-resolution 2D seismic surveys. However, the resolution and detail of the derived landscape can provide a substantive basis for further prospection or exploration of this unique landscape. In particular the need, outlined in the paper by Smith et al. (this volume), for further coring to support palaeoenvironmental research is substantially supported by our ability to identify and map deposits with enhanced environmental potential with some confidence. We may now be freed, to some extent, from our current reliance on serendipitous finds with poor contextual value. Future work should now be directed towards detailed palaeoecological studies of the type commonly carried out in terrestrial landscapes around the North Sea basin (Peeters 2006).

We should also note the potential of the data for supporting novel and exciting behavioural models with real archaeological potential (Ch'ng et al 2004). The Holocene landscapes of the North Sea were never an abstract concept. This land was both habitable and inhabited, and the landscape data we possess, or can now acquire, offers us the opportunity to explore archaeological predictive modelling that can, in turn, be used to refine our concepts of land use and enhance the potential of directed, invasive exploration to answer archaeological questions. Other research programmes have already begun to generate applied models and some, including the "Danish Fishing Model", are reported to be very successful (Fischer 1995, 375). Most of these have used localised bathymetry as a topographic proxy but this is inevitably less successful in deeper water where burial of the landscape has occurred (Fisher 1995, 377). The utilisation of information from seismic data should help improve modelling strategies and the exploration of predictive models using the North Sea seismic data is part of research currently being carried out at Birmingham (Fitch et al 2007).

Despite the apparent success of the NSPP it should not be presumed that the current work represents a final product either in spatial or chronological terms. This area studied here does not represent the whole, or even the available, extent of land surfaces that could be investigated. The shoreline of the great North Sea Holocene plain would have extended north along the current shoreline of northern England and further to the east of the present study area (Boomer et al. 2007; Coles 1998). Equally significantly, similar studies are limited by the extent of available seismic data. There is a major gap in the availability of 3D seismic data in the marine areas associated with Northern England (Bunch et al. this volume). There is also an attenuation of response to 3D seismic survey in shallow waters. Consequently, there exists a "white band" which surrounds the modern coast and within which our knowledge of the palaeotopography and, by inference, the archaeology of the area, is severely limited. Our ability to tie together the data from the Southern North Sea with terrestrial archaeology in a seamless manner, although desirable, is therefore limited. In the deeper marine areas there will be a reliance upon 2D seismics to fill this gap, with a concomitant loss of the extensive data associated with 3D data sets. In shallow waters traditional methods of marine prospection may be employed to effect (diving, high resolution seismic survey etc.): although the extent of such data is a limiting factor when considering heritage requirements to manage the resource. There is, therefore, an urgent need to collect new data sets to fill these gaps or to investigate methods to integrate other data sets in a more imaginative and productive manner.

There is another point to be made in relation to the remaining archaeological potential of the North Sea. There is increasing evidence that we should expect, at the least, low-level occupation in areas north of the study area during the Later Palaeolithic. Wickham-Jones and Dawson (2006, 19) suggest that the melting of the Devensian ice sheet north of Scotland would have been rapidly followed by marine inundation and that the areas to the north and west would not have been available for occupation. However, whilst excluding consideration of even earlier periods, the spatial extent of surviving late Palaeolithic land surfaces that have a potential for preserving traces of human occupation is actually bounded by the Norwegian Trough and encompasses the Viking Bergen Hills.

Not surprisingly, traces of occupation in these areas are few and the context of a single worked lithic, recovered from a vibrocore at a depth of 143 metres off the Viking Bank, remains uncertain (Long et al. 1986). However, Wickham-Jones and Dawson conclude that current absence of evidence for Later Palaeolithic habitation in Scotland should not be regarded as evidence of absence and that "the submerged landscape of the Scottish shelf is thus the most likely location for the preservation of traces of early settlement" (2006, 34). The significance of 3D seismic data sets from northern waters can be demonstrated and the result of analysis of one small area, not far from the Viking Bank find indicates that new insights are possible for early landscapes in the deeper waters to the north (Figure 9.2).

Such observations are also pertinent to this study. Whilst not consistently mapped as part of this project, results indicate that Later Palaeolithic surfaces are also amenable to study within the Southern North Sea (Figure 9.3 and atlas). One must assume that the potentially, well preserved Later

Palaeolithic deposits that underlie the current study area, and which stretch far to the north, must rate as priorities for research and heritage management.

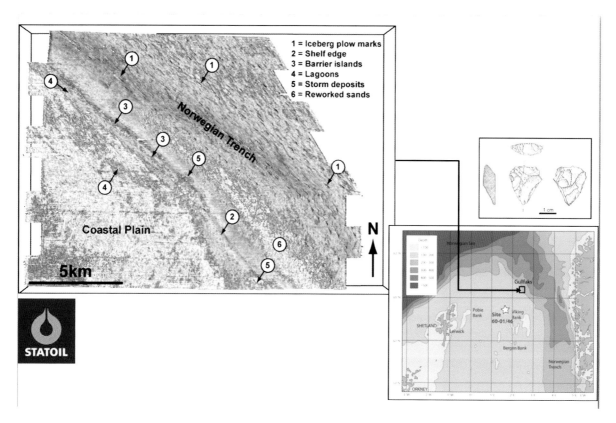

Figure 9.2 Sample seismic data illustrating probable Later Palaeolithic land surfaces adjacent to the Norwegian Trench. The Viking Bank flint illustration is from Long et al. (1986)

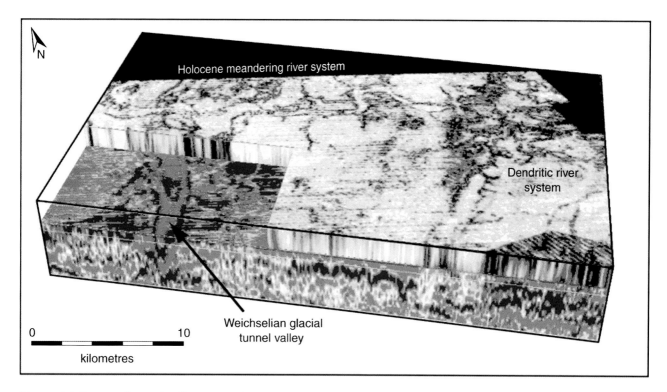

Figure 9.3 Seismic data cube illustrating chronostratigraphic relationship between Holocene and earlier features

9.3 Cultural resource management procedures in the Southern North Sea

The primary qualities of the North Sea archaeological resource, in management terms, are its general inaccessibility and an uncertainty concerning the nature, or even location of any archaeological remains. This contrasts sharply with intertidal or shallow marine zones where there is usually some opportunity to physically record, known sites, to analyse their distributions and therefore to provide some degree of protection or management. The depth of deposits, or water column, overlying the presumed North Sea landscape has generally ensured that the presence of archaeological deposits could only be inferred on the basis of contemporary correlates from terrestrial or shallow water contexts (Flemming 2004). Paradoxically, whilst there is a general assumption that the depth of water and overlying deposits might mask substantial, preserved archaeological deposits, the archaeological material trawled from the area, which is generally our only guide to the distribution of deposits, presumably suggests continuing damage to relict deposits. Unfortunately, whilst acting as an important proxy for direct examination, this material also has a low locational or interpretative value.

Management of such a resource, essentially indefinable or without adequate positional information, is an unenviable challenge but one that cannot be avoided (Roberts and Trow 2002; Oxley and O'Regan 2001). There are a number of legal or treaty obligations that govern regional and national responsibilities for marine heritage. Wickham Jones and Dawson (2007, 7-14) and Flemming (2005, 3-10) provide substantive reviews of national legislation and international obligations that apply or impact upon British maritime territory (Wenban-Smith, 2002). What need be stressed here is that whilst English Heritage's direct responsibility for marine heritage carries only to the 12 mile limit around the coast, many government agencies retain a wider interest in the marine heritage (Oxley nd). Moreover, the nation, through treaty or international obligation, is often required to consider marine heritage issues across territorial waters. It is also true, following the extensive recent activity related to marine archaeology (frequently related to ALSF funding), that agency interests in the wider issues of the North Sea are becoming more explicit (Oxley nd). This much is clear in published reviews of the potential of the marine resource (e.g. Dix et al. 2004), or projects with applied methodological value (Bates et al. nd).

A number of ALSF projects have also studied specific heritage management issues including the provision of codes of practice for reporting marine finds and investigating the application of historic landscape characterisation programmes to marine seascapes (Wessex Archaeology 2003; 2005). The establishment of the North Sea Prehistory Newsletter by Dr Peter Murphy (English Heritage) is also worth stressing here as this indicates the emergence of an international group with explicit interests in policy issues related to the prehistory of the North Sea and its associated coastline as a whole. In support of such initiatives, the overt inclusion of marine issues in the emerging Palaeolithic Research Framework[1] may also prove significant over the longer term.

Within this larger context, it is clear that the extent and detail of information provided through this project for the Holocene land surfaces of the North Sea is currently unique and tasks heritage agencies with providing an appropriate management response. Within England, at least, there are some basic guidelines to guide action, Robert and Trow's (2002) publication "Taking to the Water: English Heritage's Initial Policy for the management of Maritime Archaeology in England" sets out the general principle that the marine resource "and terrestrial archaeological remains provide a seamless physical and intellectual continuum" and that maritime heritage should "enjoy parity of esteem and treatment with their terrestrial counterparts" (Robert and Trow 2002, 4). Of particular importance within this document is the requirement to consider the marine environment as an historic <u>landscape</u>. Given the scale of the North Sea study area, this is clearly a premise from which we can begin to provide a management response (Oxley and O'Regan 2001).

With this in mind, and given the available data provided through the project, there are three heritage products that one might anticipate from this study:

- A general characterisation and interpretation of the available data in landscape terms
- An assessment for the likely potential of the available data in respect of archaeological research
- An assessment of the reliability of the interpretation and its value for mitigation mapping

The general process by which such work could be carried out as part of the project is provided in Figure 9.4.

9.4 Landscape Characterisation

For the past 16 years landscape heritage management within much of the United Kingdom has been concerned with historic landscape characterisation (HLC). HLC, in its current form, derives from the recognition that there is a requirement to provide a comprehensive characterisation of entire landscapes as a management tool (Aldred and Fairclough 2002, Fairclough and Rippon 2002). Previous heritage strategies were undermined by an overemphasis on known archaeological sites that, unintentionally, tended to privilege isolated areas to the detriment of of the wider landscape.

---

[1] http://www.iceage.org.uk/Framework.html

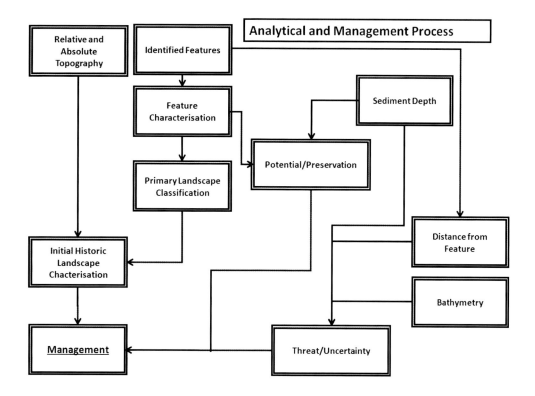

Figure 9.4 The analytical process

In contrast, HLC sought to appreciate the overall importance of the landscape itself. No specific part of any landscape is deemed as more valuable than another within an HLC programme, and all aspects of a landscape are available for classification and therefore management. Landscape is treated as a contemporary and dynamic entity incorporating past activity as one contributing factor to the final classification.

All HLC projects seek to assess every significant land parcel of a study area, assigning it to the type which best represents its predominant landscape character, as far as this is determinable. Once detailed HLC types have been defined, simplified and interpreted cultural landscape zones can be derived. This approach enables general patterns of the landscape to be discerned and provides a basis for further classification or research if required. The primary output from a traditional HLC programme is therefore a broad-brush interpretation of landscape character that supports the management of change across an entire landscape by providing a continuous assessment of the whole area (Fairclough 2006).

Heritage practitioners have not ignored the potential value of HLC within the context of maritime archaeology and a number of initiatives, including English Heritage's historic seascapes programme, have attempted to implement the broad concepts of historic landscape characterisation in a number of marine environments (Wessex Archaeology 2005; Hill et al. 2001). At the time of writing, few of these projects had reached fruition, but informal discussion with other project staff indicated that there were significant difficulties in applying HLC in areas where early features were obscured by later sediments (Baker et al. 2007 figure 8.1). In contrast, the scale at which the North Sea project has operated, and the extensive topographic data generated, has a clear HLC application.

There are, however, also significant differences faced when attempting to apply HLC to the NSPP data. In the first instance the available mapping is not consistent across the whole of the study area. Seismic response is variable and the response less good in the southern and western sectors of the study area where the water column is relatively shallow. The resolution of the mapped data from the North Sea also falls well below that expected by terrestrial HLC projects, which commonly use 1:1250-1:10,000 map scales according to the context of the project (Aldred and Fairclough 2002, 26). However, the poor resolution of the data may be offset by the nature of the landscape under study. The North Sea data effectively represents a partially mapped Mesolithic landscape in topographic terms, and the notional resolution of the data (Thomson and Gaffney, this volume), supports mapping of generalised economic/landscape units which may, ultimately, reflect broad land use patterns within a Mesolithic economy. Whilst unencumbered by later cultural development the landscape's post-depositional taphonomy should also be considered part of the landscape's character. In this sense, the data seems, *a priori*, to possess the potential for a successful HLC implementation.

Table 9.1 Primary landscape characterisation zones

Class	Area (Km^2)	Logical extension/ Speculative (Km^2)	ID	Description	General Areas
1	1872.33	261.63	Early Mesolithic Seaway	Similar to the Severn Estuary, base shows tidal scour marks and presence of abandoned bedforms.	The Outer Silver Pit
2	2872.34	0	Dominated by geology with fluvial systems	Area is typified by thin deposits and near surface solid geology, which illustrate modern erosion, a few fragmentary fluvial systems present.	Offshore Lincolnshire
3	3154.09	932.17	Dominated by geology with some fluvial systems	Area of very strong solid geology response with large salt structures. It is likely that these formed regional features within the landscape.	Spurn, Sole Pit Region and Eastermost Rough
4	412.55	157.45	Inlet area - partial scour	The inlet of the Outer Silver Pit. The area only shows partial scouring, and possesses a channel that may have drained the lake which once existed in the Outer Silver Pit.	Outer Silver Pit
6	4390.82	124.84	Landscape influenced by underlying glacial deposit	An area where the fluvial channels appear to be influenced by underlying glacial deposits.	Offshore Norfolk, South Central Dogger Bank, Outer Well bank
7	3760.9	0	Area of smaller Holocene channels	This area contains many smaller channels, only visible in part, but appear to have extended across the whole of this area.	Swarte Bank, Indefatigable Bank
8	823.52	0	Area surrounding large lacustrine feature	Drainage in this region appears to have been dominated by the Markhams Hole Lacustrine System.	South Botney Cut, South East Outer Silver Pit
9	2762.3	65.54	Low lying areas with soft coastline	Lower lying areas of the Doggerbank region that possess extensive fluvial systems which extend into soft coastline areas.	Southern Doggerbank, South West Patch and South West Spit
10	1626.13	0	Area of reuse of Later Pleistocene features	This area is dominated by two major fluvial systems which appear to be utilising existing Late Pleistocene courses.	South Eastern Outer Silver Pit, Well Hole
11	146.37	0	Lacustrine feature	Area that would have formed lakes/wetlands during the Early Mesolithic.	Sole Pit, Silver Pit, Well Hole and Markhams Hole
12	1033.29	0	Doggerbank fluvial systems area	This area holds the clearest and best preserved of the fluvial systems in the region, the data in these regions suggest that the landscape is well preserved.	Dogger bank, Eastermost Shoal
13	535.23	0	Areas with clear indication of marine transgression (salt marsh)	These areas clearly show evidence of being altered by marine incursion.	Botney Cut Region, Cleaver Bank, South Rough, Eastermost Shoal
14	293.23	42.67	Early Holocene coastline	The Early Mesolithic coastline.	Outer Silver Pit, North East Doggerbank

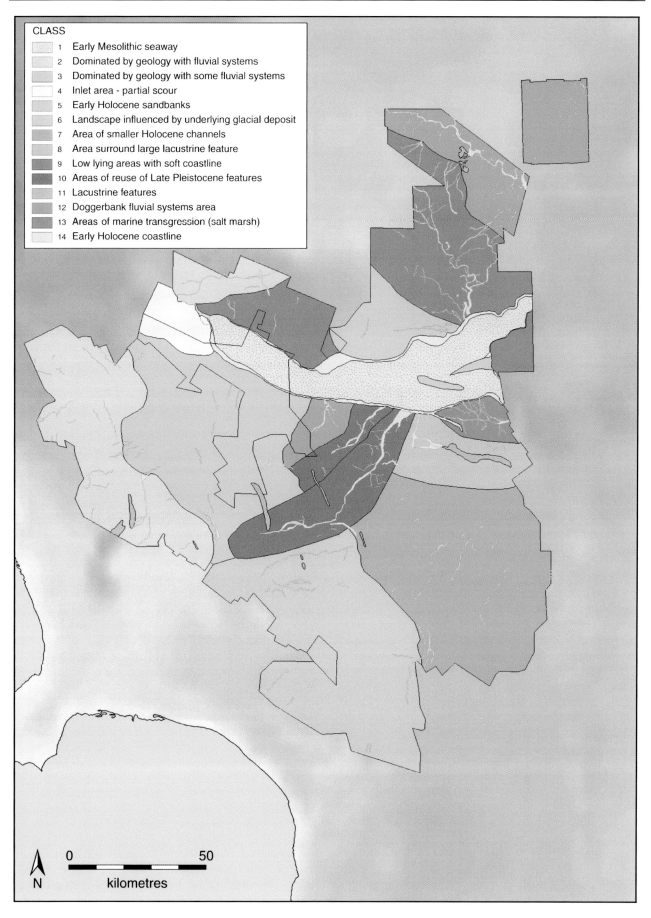

Figure 9.5 Broad landscape character zones

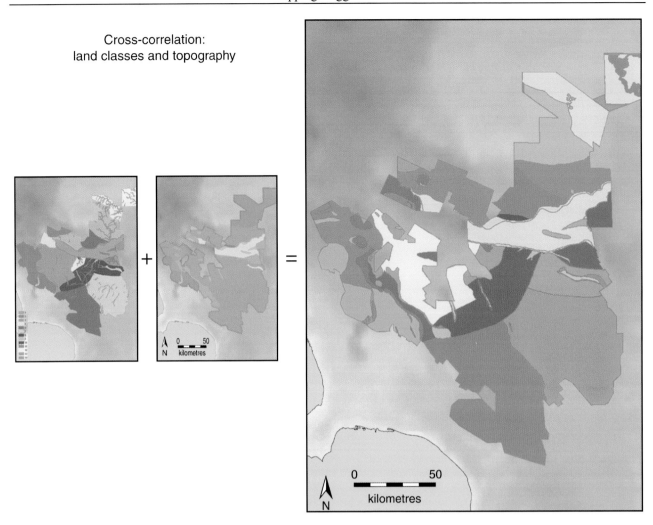

Figure 9.6 Cross correlation of major topographic and landscape characterisation zones

On that basis, a strategy was designed to provide a basic HLC product using the available data. Initially, the landscape was classified into a series of broad areas based upon their depositional history, and major historic landscape features. At the outset an attempt was made to automate this process by treating the data as a hyperspectral image. This followed a procedure designed to categorise poorly mapped landscapes explored by the team in the context of research at Fort Hood, in Central Texas (Barratt et al 2007; White and Ray 2000). This approach, however, failed at an early stage due to the mosaic nature of the seismic images. The landscape characterisation was therefore performed manually using the available mapping, and was primarily guided by geomorphological and hydrological characteristics to provide broad landscape zones. On this basis, the entire area was classified into the areas detailed in Table 9.1, and the description added to the polygon layer as an attribute to provide graphical display (Figure 9.5). The dividing lines for many of the landscape zones observed coincided broadly with known watersheds between observed fluvial features.

However, provision of the broad topographic variation of the landscape, picked from the Holocene land surfaces, permitted refinement of this image by cross-tabulating the primary topographic variation of the Mesolithic landscape with the primary landscape zones defined from the initial characterisation phase. A total of 80 separate land classes were generated through this process and the result is shown in Figure 9.6. This data is interesting as it probably represents the best general zonation, in terms of potential Holocene land use, currently achievable using the available data and, to the extent that it may correlate with broad economic activity, may carry considerable potential to act as the basis for more detailed behavioural modelling.

It should be acknowledged that the data presented here does not currently represent a full HLC product, as it does not truly incorporate contemporary landscape features. In fact a further reclassification of the primary HLC image was generated incorporating a zoned bathymetry layer. This produced an excessively complex image that, although potentially of use for management, is not presented here and is retained in archive.

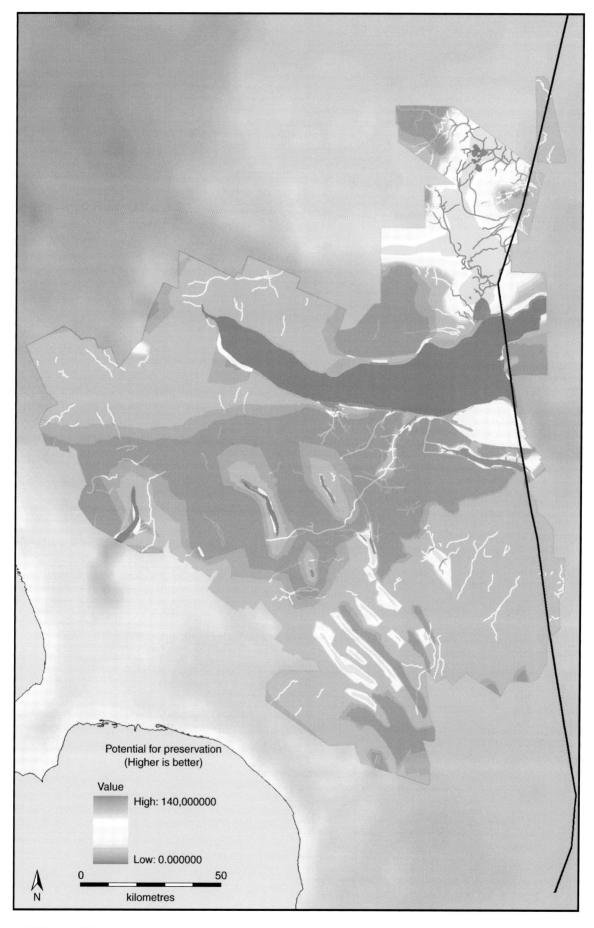

Figure 9.7 Potential for preservation

Table 9.2 Ranking of features by relative archaeological potential

River potential 0 = Absent 1 = Low 4 = high (Ranking based on a modified Strahler stream ordering)	Lakes 0 = Absent 4 = present	Marsh/Wetlands 0 = Absent 4 = present	Coastlines 0 = Absent 4 = Present	Deadzones/Background Scoured = 0 Landscape present = 1

9.5 Threat mapping

Whilst the HLC data are significant in their own right they are probably not, in themselves, an adequate basis for a larger management strategy for the southern North Sea. In particular, this requires an assessment of the potential of the area under study for preservation of archaeological materials. This may be provided by the multiplication of two normalised data sets for the interpreted archaeological potential of identified landscape features and the depth of overlying sediments derived from published sources including BGS mapping. The relative values for landscape feature potential, prior to normalisation, are provided in Table 9.2. Following this process, areas with a lack of known features and an absence of significant sedimentation score low, whilst probable scoured areas, including the Outer Silver Pit, produce a value of 0. Areas with probable archaeological potential and with significant overlying deposits score high. The mapped data is provided in Figure 9.7.

Not surprisingly, Figure 9.7 emphasises areas that are likely to be of prime archaeological interest, most notably lacustrine environments, marsh areas and coasts. However, the areas around the large river systems to the north of the Outer Silver Pit are emphasised overall; a consequence of the association of a dense network of major channels and protective sediments. The apparent potential associated with the large sand bank systems in the south east of the project area may be misleading as this reflects depth of sediment associated with highly mobile features.

9.6 Threat and Uncertainty Mapping

The data provided in Figure 9.7 is useful in assessing the overall significance of features identified through the seismic analysis. However, such mapping does not provide substantial guidance in areas within which features have not been identified. Hence the extensive areas that are suggested as having a relatively low potential may be misleading. Earlier papers in this volume have noted that our ability to identify Holocene structures is limited in a number of areas, notably those associated with a shallow water column. Consequently, a primary concern, after direct identification of Holocene features, must be to identify areas that may contain features and might also be under threat. Such zones may be chosen for further prospection or development plans modified in the light of the potential of such areas to contain undiscovered features.

When considering these issues it seems reasonable to suggest that the further we are from identifiable structures the more likely it is that other factors may be preventing discovery. Following such an argument, a separate map was prepared representing threat and uncertainty as a single measure linked to horizontal distance from feature and accessibility (ranked according to the overlying depth of sediment and water column). These three factors were normalised and summed to provide a single value and the results are presented in Figure 9.8. This map provides a continuous assessment across the study area in which areas of high threat and low uncertainty (shallow water column or sediments proximate to identified features) grade into areas with low threat and high uncertainty (greater water column or sediments at an increasing distance from known features).

This procedure results in a simple but highly effective form of "red flag" mapping that can be usefully compared to Figure 9.2, which primarily reflects probable archaeological potential. Setting aside areas which may be scoured (the Outer and Inner Silver Pits), this measure highlights significant areas in the southern and western parts of the study area as zones which might contain features, which are not amenable to current mapping technologies but which may be more prone to development threat.

9.7 Final Observations

The North Sea Palaeolandscapes Project mapped more than 23,000 square kilometres of Holocene land surfaces and presented these for publication in slightly over 18 months. The product of this work is one of the largest analyses of remotely sensed data ever attempted for archaeological purposes and this has brought to the attention of the archaeological and heritage communities one of the most extensive and best preserved prehistoric landscapes in Europe at least. The methodologies demonstrated here have wide application in similar landscapes elsewhere, when appropriate remote sensed data is available. However, whilst such research is technically appealing, we should not lose sight of the fact that its fundamental importance is its ability to inform research into the Early Holocene communities of northwestern Europe.

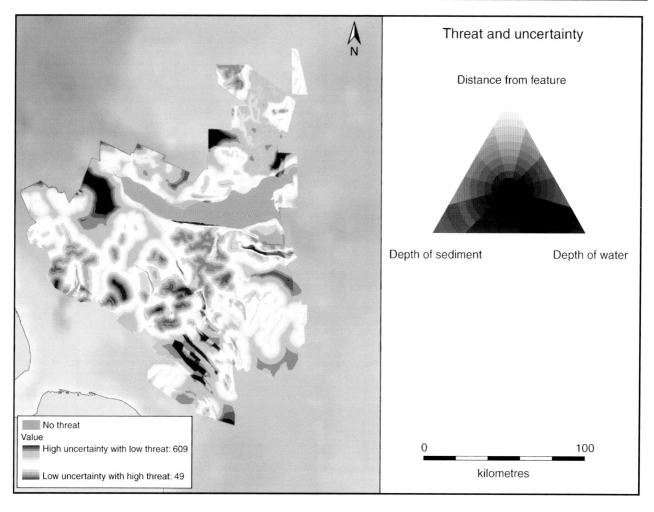

Figure 9.8 Red flag mapping. This figure combines threat and uncertainty data based on distance to feature and depth of overlying sediment. The lack of sediment cover and direct association with identified features with archaeological potential rate as high threats with little uncertainty. Deep overlying deposits lying further from recorded features rank as low threat areas but with significant levels of uncertainty.

Reynier (2005, 1) recently described research into the early Mesolithic as currently "listless", perhaps due to the difficulties presented by the archaeological record. In part this may be a consequence of our lack of knowledge of the prehistoric archaeology of the North Sea. Currently, the Holocene archaeology of the region is infrequently considered within the literature and the lack of available evidence is tacitly presented as an absence of evidence. Consequently, the area appears to occupy a proximal role in the literature and our interpretative position. It remains true, however, that only a few terrestrial sites are actually available to support our current interpretative models for the earlier Holocene (Milner and Woodman 2004, 5), and fewer have provided adequate environmental evidence for this period (Whitehouse and Smith 2004). This is a parlous position and we should be assured that the apparent density of sites that have been identified or explored in Europe, most notably in Denmark, will not actually fill this gap (Fischer 2004). Few of these sites are located further than 5km from the coast and, whilst useful for comparison, these can never truly be used as a proxy for settlement more than 120km away, in the centre of the Great North Sea plain.

The results of the NSPP should be taken as a wake-up call. The landscapes mapped here represent areas that would have been prime habitable zones linking and, perhaps, explaining much of the archaeological variation we see around the North Sea basin. In contrast, the present terrestrial archaeological record, which approximates the sum of our knowledge, in Britain at least, may better be represented as areas that were peripheral locations for the Mesolithic occupants of the North Sea basin (Morrison 1980). The data provided by the NSPP mapping programme provides significant support for a radical shift in our interpretative position for the Mesolithic in north western Europe. Previously unimaginable, the Holocene landscapes revealed here allow us to discriminate between environmental zones, characterise areas of archaeological potential and, possibly, provide the opportunity to explore the southern North Sea with the likelihood of archaeological success. In doing so we can anticipate the exploration of an entirely new European country whose study may re-invigorate research into the Mesolithic and later Palaeolithic occupation of northwestern Europe.

In promoting academic change the project can, so far, be judged a success. However, the exploration and mapping of uncharted rivers and hills also brings other responsibilities. The archaeology of the North Sea has remained essentially untouched since the land was lost to the sea more than 7,000 years ago. Whilst it is easy to be overwhelmed by the sheer scale of the surviving historic landscape, the heritage of the North Sea is fragile. Our previous lack of knowledge has permitted unsympathetic development and poorly managed exploitation. Having rediscovered "Doggerland" the United Kingdom, and all the countries bounding the North Sea, must assume the responsibility associated with our shared heritage. How we respond to such a unprecedented challenge will be judged by future generations.

References

Aldred, O., and Fairclough, G. 2003. *Historic Landscape Characterisation: taking stock of the method.* English Heritage and Somerset County Council, London.

Allen, J. R. L. 1997. Subfossil mammalian tracks (Flandrian) in the Severn Estuary, S.W. Britain: mechanics of formation, preservation and distribution. *Philosophical Transactions of the Royal Society* **B352** (513): 481-518.

Allen, M. J., and Gardiner, J. 2000. *Our Changing Coast: a survey of the intertidal archaeology of Langstone Harbour, Hampshire.* CBA Research Report **124**.

Amos, C. L. and King, E. L. 1984. Bedforms of the Canadian Eastern seaboard: a comparison with global occurrences. *Marine Geology.* **57**: 167-208.

Ashley, G. M. 1990. Classification of large-scale subaqueous bedforms: a new look at an old problem. *Journal of Sedimentary Petrology.* **60**: 160-172.

Autin, W. J. 2002. Landscape evolution of the Five Islands of south Louisiana: scientific policy and salt dome utilization and management. *Geomorphology.* **47**: 227-244.

Bacon, M., Simms, R., and Redshaw, T. 2003. *3-D Seismic Interpretation.* Cambridge University Press.

Bahorich, M., and Farmer, S. 1995. 3-D seismic discontinuity for faults and stratigraphic features: The coherence cube. *The Leading Edge* **14**: 1053-1058.

Bailey, G. 2004. The wider significance of submerged archaeological sites and their relevance to world prehistory. In Flemming N.C (ed.) *Submarine Prehistoric Archaeology of the North Sea*, CBA Research Report **141**.

Baker, M.V., Tapper, B, Johns, C., and Herring P. 2007. *England's Historic Seascapes, Scarborough to Hartlepool and Adjacent Marine Types. Historic Seascape Characterisation.* Historic Environment Service, Environment and Heritage, Cornwall County Council Report No: 2007R021.

Ballard, R. D., Colemand, D. F., and Rosenburg G. 2000. Further Evidence of Abrupt Holocene Drowning of the Black Sea Shelf. *Marine Geology.* **170**: 253-61.

Bally, A. W. 1987. Atlas of seismic stratigraphy Volume 1. *AAPG studies in Geology*: **27**. American Association of Petroleum Geologists, Tulsa, Oklahoma, U.S.A.

Balson, P. S., and Jeffery, D. H. 1991. The glacial sequence of the southern North Sea. In J.Ehlers, P. L. Gibbard, and J. Rose, (eds.) *Glacial Deposits in Great Britain and Ireland.* Rotterdam: Balkena, 245-253.

Barratt, G., Dingwall, L., Gaffney, V., Fitch, S., Huckerby, C., and Maguire, T. 2007. *Heritage management at Fort Hood, Texas: Experiments in Historic Landscape Characterisation.* Archaeopress.

Barton, N., and Roberts, A. 2004. The Mesolithic Period in England: Current perspectives and new research. In: A. Saville (ed.) *Mesolithic Scotland and its Neighbours.* Society of Antiquaries of Scotland.

Bates, R., Dean, M., Lawrence, M., Robertson, P., and Tempera, F. 2005. *Innovative Approaches to Archaeological Site surveying and Evaluation.* ALSF Project 3837. http://www.coastms.co.uk/Conferences/Outputs%20and%20 Reports/ALSF%202005/ALSF%202005%20Bates.pdf

Bell, M., Caseldine, A., and Neumann, H. 2000. *Prehistoric Intertidal Archaeology in the Welsh Severn Estuary* (CBA Research Report 120). London: Council for British Archaeology.

Bell, M., and Walker, M.J.C. 2005. *Late Quaternary Environmental Change* (second edition). Harlow: Pearson.

Bell, T., O' Sullivan, A., and Quinn, R. 2006. Discovering ancient landscapes under the sea. *Archaeology Ireland* **20** (2): 12-17.

Bellamy, A. G. 1998. The UK marine sand and gravel dredging industry: an application of Quaternary geology. In: J. - P. Latham, (ed.) *Advances in Aggregates and Armourstone Evaluation.* Geological Society, London, Engineering Special Publications, **13**: 33-45.

BGS. 2007. - Seismicity and Earthquake Hazard in the UK - located at http://www.earthquakes.bgs.ac.uk/hazard/Hazard_UK.htm - visited 14/02/2007.

Bird, E. 2000. *Coastal Geomorphology: An Introduction.* John Wiley and Sons, 322.

Bjerk, H. B. 1995. The North Sea Continent and the pioneer settlement of Norway. *In*. A. Fischer (ed.) *Man and Sea in the Mesolithic.* Oxbow. 131- 144.

Blondeaux, P. 2001. Mechanics of coastal forms. *Annual Review of Fluid Mechanics.* **33**: 339-370.

Boomer, I., Waddington, C., Stevenson, T., and Hamilton, D. 2007. Holocene coastal change and geoarchaeology at

Howick, Northumberland, UK. *The Holocene* **17** (1): 89-104.

Brayshay, B. A. and Dinnin, M. 1999. Integrated palaeoecological evidence for biodiversity at the floodplain-forest margin. *Journal of Biogeography* **26** (1): 115-131.

Brown, A. G. 1983. Floodplain deposits and accelerated sedimentation in the lower Severn basin in K. J. Gregory (ed) *Background to Palaeohydrology.* London: Wiley. 375-397.

Brown, A. 1997. *Alluvial Archaeology.* Cambridge: Cambridge University Press.

Brown, M. 1986. *Indefatigable Sheet (53N 02E): Quaternary Geology.* HMSO, London

Bruthans, J., Filippi, M., Geršl, M., Zare, M., Melková, J., Pazdur, A., and Bosák, P. 2006. Holocene marine terraces on two salt diapirs in the Persian Gulf, Iran: age, depositional history and uplift rates. *Journal of Quaternary Science.*

Bulat, J. 2005. Some considerations on the interpretation of seabed images based on commercial 3D seismic in the Faroe-Shetland Channel. *Basin Research* **17**: 21-42.

Bull, J. M., Gutowski, M., Dix, J. K., Henstock, T. J., Hogarth, P., Leighton, G., and White, P. R. 2005. *Marine Geophysical Researches* **26**: 157-169.

Burkitt, M. C. 1932. A Maglemose Harpoon dredged up recently from the North Sea. *Man* **32** (May, 1932): 118.

Buteux, S., Gaffney, V., van Leusen, M., and White, R. 1997. Establishing GIS services at the Birmingham University Field Archaeology Unit: Problems and potential. In A. Gotarelli (ed) *Sistemi Informativi e Reti Geografiche in Archeologia: GIS-Internet.* Firenze, 59-77.

Buteux, S., Gaffney, V., White, R., and van Leusen, M. 2000. Wroxeter Hinterland Project and geophysical survey at Wroxeter. *Archaeological Prospection* **7**: 69-80.

Cameron, T. D. J., Crosby, A., Balson, P. S., Jeffery, D. H., Lott, G. K., Bulat, J., and Harrison, D. J. 1992. *United Kingdom offshore regional report: the geology of the Southern North Sea.* HMSO for the British Geological Survey.

Carr, S. J., Holmes, R., Van Der Meer, J. J. M., and Rose, J. 2006. The Last Glacial Maximum in the North Sea Basin: micromorphological evidence of extensive glaciation. *Journal of Quaternary Science.* **21**: 131-153.

Ch'ng, E., Stone, R. J., and Arvantis, T. N. 2004. The Shotton River and Mesolithic Dwellings: Recreating the Past from Geo-Seismic Data Sources. *The 5th International Symposium on Virtual Reality, Archaeology and Cultural Heritage*, VAST (2004), in cooperation with ACM SIGRAPH and The Eurographics Association, 7-10 December, Brussels, Belgium. 125-133.

Chapman, H. P. 2006. *Landscape Archaeology and GIS.* Tempus Publishing Ltd. 60.

Chapman, H. P., Fletcher, W. G. and Thomas, G. 2001. Quantifying the effects of erosion on archaeology of inter-tidal environments: a new approach and its implications for their management. *Conservation and Management of Archaeological Sites* **4**: 233-240.

Chatterton, R. 2005. Ritual. *In.* C. Conneller and G. Warren. *Mesolithic Britain and Ireland: New Approaches.* Tempus.

Clarke, G. 1936. *The Mesolithic Settlement of Europe.* Cambridge University Press.

Clark, J. G. D. 1972. Star Carr: a Case Study in Bioarchaeology. Reading (MA): Addison-Wesley (Addison Module in Anthropology 10).

Clark, J. G. D. 1975. The Earlier Stone Age Settlement of Scandinavia. Cambridge University Press.

Clarke, D. 1978. *Mesolithic Europe: The Economic Basis.* Duckworth.

Coles, B. J. 1998. *Doggerland: a speculative survey.* Proceedings of the Prehistoric Society **64**: 45-81.

Coles, B. J. 1999. Doggerland's loss and the Neolithic. In: B. Coles, J. Coles and M. Schon Jorgensen (eds.) *Bog Bodies, Sacred Sites and Wetland Archaeology.* WARP Occasional Paper. **12**: 51-57

Coles, B. J. 1998. Doggerland: a speculative survey. *Proceedings of the Prehistoric Society.* **64**: 45-81.

Connolly, J., and Lake, M. 2006. *Geographical Information Systems in Archaeology.* Cambridge Manuals in Archaeology, Cambridge University Press.

Cook P.J. 1991. Spurn Sheet *(53N 00E): Quaternary Geology.* HMSO, London

Darvill, T. C. 1995. *Prehistoric Britain.* Routledge

Davies, J. L. 1964. A morphogenetic approach to world shorelines. *Zeitschrift fur Geomorphologie.* **8**: 127-142.

Davison, I., Bosence, D.W., Alsop, G.I. & Alawi, M. 1996. Deformation and sedimentation around active Miocene salt diapirs on the Tihama Plain, northwest Yemen. In: G. I.

Alsop, D. J. Blundell, and I. Davison, (eds.) *Salt Tectonics*. Geological Society, London, Special Publications. **100**: 23-39.

Dawson M. 2000 (ed). *Prehistoric, Roman and post-Roman landscapes of the Great Ouse Valley* (Council for British Archaeology Research Report 119) York: Council for British Archaeology.

De Serres, B., Roy, A. G., Biron, P. M., and Best, J. L. 1999. Three-dimensional structure of flow at a confluence of river channels with discordant beds. *Geomorphology.* **26**: 313-335.

Delgado, J. 1998. Encyclopedia Of Underwater And Maritime Archaeology. Yale University Press.

Den Hartog Jager, D., Giles, M. R., and Griffiths, G. R. 1993. Evolution of Palaeogene submarine fans of the North Sea in space and time. In: J. R. Parker, (ed) *Petroleum geology of Northwest Europe: Proceedings of the 4th Conference. Geological Society of London*, Bath, 59-72.

Dix, J., Quinn, R., and Westley, K. 2004. *A Reassessment of the Archaeological Potential of Continental Shelves*. Located at: http://www.arch.soton.ac.uk/Research/Aggregates/shelve-report.htm

Donovan, D. T. 1965. Holocene origin of erosion hollows in the North Sea. *Challenger Society Annual Report.* **3** (17): 43-44.

Donovan, D. T. 1975. The Geology and origin of the Silver Pit and other closed basins in the North Sea. *Proceedings of the Yorkshire Geological Society.* 39 (2): 267-293.

Dortch, C. 1997. Prehistory Down Under: archaeological investigations of submerged Aboriginal sites at Lake Jasper, Western Australia. *Antiquity* **71** (271): 116-123.

Dyer, K. R., and Huntley, D. A. 1999. The origin, classification and modelling of sand banks and ridges. *Continental Shelf Research.* **19**: 1285-1330.

Eisma, D. 1975. Holocene sedimentation in the Outer Silver Pit Area (Southern North Sea). *Marine Science Communications.* **1**: 407-426.

Emery, D. and Myers, K.J. 1996. *Sequence Stratigraphy*. Blackwell Science.

English Heritage. (nd). (http://www.english-heritage.org.uk/server/show/nav.1315)

Fairclough, G. J. 2006. From assessment to characterisation. *In* J. Hunter, and I. Ralston, (eds.) *Archaeological Resource Management in the UK*, Second Edition. Sutton. 250-270.

Fairclough G and Rippon S. 2002. *Europe's Cultural Heritage: archaeologists and the management of change.* Europae Archaeoloiae Consilium. Brussels.

Faught, M. K. 1988. Inundated Sites in the Apalachee Bay area of the Eastern Gulf of Mexico. *Florida Anthropologist.* **41**: 185-190.

Faught, M. K. 2004. The Underwater Archaeology of Palaeolandscapes, Apalachee Bay, Florida. *American Antiquity,* **69** (2): 275-289.

Fenies, H., de Resseguier, A., and Tastet, J.-P. 1999. Intertidal clay-drape couplets (Gironde estuary, France). *Sedimentology.* **46**: 1-15.

Fischer, A. 1995. An Entrance to the Mesolithic world beneath the Ocean. Status of ten years' work on the Danish sea floor. *In* A. Fischer (ed) *Man and Sea in the Mesolithic: Coastal settlement above and below present sea level*, Proceedings of the International Symposium, Kalundborg, Denmark 1993. Oxbow.

Fischer, A. 2004. Submerged Stone Age: Danish examples and North Sea potential. In: N. C. Flemming (ed.) *Submarine Prehistoric Archaeology of the North Sea.* CBA Research Report **141**: 21-36.

Fitch, S., Thomson, K., and Gaffney, V. L. 2005. Late Pleistocene and Holocene depositional systems and the palaeogeography of the Dogger Bank, North Sea. *Quaternary Research* **64**: 185-196.

Fitch, S., Gaffney, V. L., and Thomson, K. 2007. In Sight of Doggerland: From speculative survey to landscape exploration. *Internet Archaeology* 22. (http://intarch.ac.uk).

Flemming, N. C. 2002. *The scope of Strategic Environmental Assessment of North Sea areas SEA3 and SEA2 in regard to prehistoric archaeological remains.* Department of Trade and Industry. Report TR_014.

Flemming, N. C. 2003. *The scope of Strategic Environmental Assessment of Continental Shelf Area SEA4 in regard to prehistoric archaeological remains.* Department of Trade and Industry.

Flemming N. C. 2004. *Submarine Prehistoric Archaeology of the North Sea.* CBA Research Report **141**.

Flemming, N. C. 2004b. *The scope of Strategic Environmental Assessment of North Sea areas SEA5 in regard to prehistoric archaeological remains.* Department of Trade and Industry.

Flemming, N. C. 2005. *The scope of Strategic Environmental Assessment of Irish Sea Area SEA6 in regard to prehistoric archaeological remains.* Department of Trade and Industry Report.

Flemming N. C. 2006. *Area SEA6 with regard to prehistoric and early historic archaeological remains. Department of Trade and Industry.* (http://www.offshore-sea.org.uk/site/scripts/category_info.php?categoryID=37)

Friedrichs, C. T., and Aubrey, D. G. 1988. Non-linear tidal distortion in shallow well mixed estuaries: a synthesis. *Estuarine and Coastal Shelf Science.* **27**: 521-546.

Fuglestveldt, I. 2003. Encultring the Landscape beyond Doggerland. *In* L. Larsson, H. Kindgren, K. Knutsson, D. Leoffler, and A. Åkerlund (eds.). *Mesolithic on the Move: Papers presented at the Sixth International Conference on the Mesolithic in Europe, Stockholm 2000.* Oxbow. 103-106.

Funnel, B. M. 1995. Global sea-level and the (pen-) insularity of late Cenozoic Britain. *In* R. C. Preece (ed.) *Island Britain: a Quaternary Perspective,* 3 -13. London: Geological Society.

Gaddy, D. E. 2003. *Introduction to GIS for the Petroleum Industry.* Penwell publishers.

Gaffney, C., and Gaffney, V. 2000a. Non-invasive Investigations at Wroxeter at the end of the 20th Century. *Archaeological Prospection* 7(2).

Gaffney, C., Gater, J., Linford P., Gaffney V., and White R. 2000b. Large-scale systematic fluxgate gradio-metre at the Roman city of Wroxeter. *Archaeological Prospection* **7**: 81-99.

Gaffney, C., and Gater, J. 2003. *Revealing the Buried Past: Geophysics for Archaeologists.* Tempus Publishers.

Glimmerveen J., Mol D., Post K., Reumer J. W. F., van der Plicht, H., de Vos, J., van Geel, B., van Reenen, and Pals, J. P. 2004. The North Sea project: the first palaeontological, palynological and archaeological results. In: N. C. Flemming 2004 (ed.). *Submarine Prehistoric Archaeology of the North Sea.* CBA Research Report **141**: 21-36.

Godwin H and Godwin M. E. 1933. British Maglemose harpoon sites. *Antiquity* **VII** (25): 36-48.

Goodman, D., and Nishimura, Y. 2000. Ground-penetrating Radar Survey at Wroxeter. *Archaeological Prospection,* 7(2): 101-107.

Greenwood, M., and Smith, D. N. 2005. A survey of Coleoptera from sedimentary deposits from the Trent Valley, in D. N. Smith, M. B. Brickley and W. Smith (eds). *Fertile Ground: Papers in Honour of Professor Susan Limbrey* (AEA Symposia no. 22). Oxford: Oxbow Books 53-67.

Grimm, E. 1991. *TILIA and TILIA.GRAPH.* Illinois State Museum.

Groom, D., and Oxley, I. 2001. Maritime Fife: managing Fife's underwater heritage: a feasibility study for a maritime archaeological GIS. In D. W. Wheatley, G. P. Earl and S. J. Poppy (eds.) *Contemporary Themes in Archaeological Computing.* Oxbow Books.

Gutowski, M. 2005. *3D high-resolution sub-bottom profiling - 3D Chirp.* Hydro International.

Gutowski, M., Bull, J., Dix, J., Henstock, T., Hogarth, P., Leighton, T., and White, P. 2005. True 3D high resolution sub-bottom profiling: 3D Chirp system design and data examples. *Geophysical Research Abstracts,* **7**: No.06616, EGU.

Harris, P. T. 1988. Large-scale bedforms as indicators of mutually evasive sand transport and the sequential infilling of wide-mouthed estuaries. *Sedimentary Geology.* **57**: 273-298.

Harris, P. T., Pattiaratchi, C. B., Cole, A. R., and Keene, J. B. 1992. Evolution of subtidal sandbanks in Moreton Bay, eastern Australia. *Marine Geology.* **103**: 225-247.

Haslett, S. K. 2000. *Coastal Systems.* Routledge. 218.

Hayes, M. O. 1975. Morphology of sand accumulation in estuaries: an introduction to the symposium. In L. E. Cronin (ed). *Estuarine Research. Volume II: Geology and Engineering.* Academic Press, New York. 3-22.

Healy, F., Heaton, M., and Lobb, S. J. 1992. Excavation of a Mesolithic site at Thatcham, Berkshire. *Proceedings of the Prehistoric Society.* **58**: 41-76.

Hill, M., Brigg, J., Minto, P., Bagnal, D., Foley, K., and Williams, A. 2001. *Guide to Best Practice in Seascape Assessment.* Countryside Council for Wales. Brady Shipman Martin, University College Dublin. http://www.ccw.gov.uk/Images_Client/Reports/ACF1676.pdf.

Horton, B. P., Innes, J. B., Shennan, I., Gehrels, W. R., Lloyd, J. M., McArthur, J. J. and Rutherford, M. M. 1999. The Northumberland Coast in D. R. Bridgland, B. P. Horton and J. B. Innes (eds) *The Quaternary of North-East England. Field Guide.* London: Quaternary Research Association. 147-165.

Howard, A. J. 2005. The contribution of geoarchaeology to understanding the environmental history and archaeological resources of the Trent Valley, U.K. *Geoarchaeology* **20**, 93-107.

Howard, A. J., and Macklin, M. G. 1999. A generic geomorphological approach to archaeological interpretation and prospection in British River valleys: A guide for archaeologists' investigation of Holocene Landscapes. *Antiquity* **73**, 527-41.

Huntley, B., and Birks, H. J. B. 1983. *An atlas of past and present pollen maps of Europe, 0-13,000 years BP.* Cambridge University Press.

Indrelid, S. 1978. Mesolithic economy and settlement patterns in Norway. In P. A. Mellars (ed.) *The early postglacial settlement of Northern Europe: An ecological perspective.* Duckworth. 147-76.

Jackson, M. P. A., and Talbot, C. J. 1994. Advances in Salt Tectonics. In: P. L. Hancock, (ed.) *Continental Deformation.* Pergamon Press, 159-179.

Jacobi, R. M. 1973. Aspects of the Mesolithic Age in Great Britain. In S. K. Kozlowski (ed.) *The Mesolithic in Europe.* Warsaw University Press. 237-266.

Jacobi, R. 1976. Britain inside and outside Mesolithic Europe *Proceedings of the Prehistoric Society* **42**, 67-84.

Jacobi, R.M. 1978. The Mesolithic of Sussex. In. P.L Drewett (ed.) Archaeology in Sussex to AD 1500. Council for British Archaeology Research Report, 29, 15-22.

Jelgersma, S. 1979. Sea-level changes in the North Sea basin. In: E. Oele, R. T. E. Schuttenhelm, and A. J. Wiggers, (eds) *The Quaternary history of the North Sea.* Acta Universitatis Upsaliensis: Symposium Universitatis Upsaliensis Annum Quingentesimum Celebrantis: **2**: 233-248.

Jenyon, M. K. 1986. *Salt Tectonics.* Elsevier, London.

Johnson, H. D., and Baldwin, C. T. 1996. Shallow Clastic Seas. In H. G. Reading (ed.) *Sedimentary Environments: Processes, Facies and Stratigraphy.* Blackwell Science. 232-280.

Kenward, H. K., Engleman, C., Robertson, A., and Large, F. 1986. Rapid scanning of urban archaeological deposits for insect remains. *Circaea* **3**, 163-172.

Kenward, H. K., and Hall, A. R. 2006. Easily decayed organic remains in urban archaeological deposits: value, threats, research directions and conservation, in O. Brinkkemper, J. Deeben, J. van Doesburg, D. Hallewas, E. M. Theunissen, and A. D. Verlinde, (eds), *Vakken in vlakken: Archeologische kennis in lagen.* (Nederlandse Archeologische Rapporten 32). Amersfoort: Rijksdienst voor het Oudheidkundig Bodemonderzoek. 183-198.

Kenyon, N. H. 1970. Sand ribbons of European tidal seas. *Marine Geology.* **9**: 25-39.

Kenyon, N. H., Belderson, R. H., Stride, A. H., and Johnson, M. A. 1981. Offshore tidal sandbanks as indicators of net sand transport and as potential deposits. *International Association of Sedimentologists, Special Publication.* **5**: 257-268.

Kidd, G. D. 1999. Fundamentals of 3D seismic volume visualization. *The Leading Edge* **18**: 702-709.

Klemperer, S., and Hobbs, R. 1991. *The BIRPS atlas: deep seismic reflection profiles around the British Isles.* Cambridge University Press.

Knight, D., and Howard. A. J. 2005. *Trent Valley Landscapes.* Heritage, Marketing and Publications Ltd.

Knighton, D. 1998. *Fluvial Forms and Processes: A New Perspective.* Arnold. 383.

Kooijmans, L. P. 1971. Mesolithic bone and antler implements from the North Sea and from the Netherlands. *Berichtien van de Rijksdienst voor het Oudheidkunde Bodemonderzoek*, **20-21**: 27-73.

Kvamme, K. 2006. Integrating multidimensional geophysical data. *Archaeological Prospection* **13**: 57-72.

Lambeck, K. 1995. Predicted Shoreline from Rebound Models. *Journal of the Geological Society* **152**: 437-448.

Lambeck K. 1996. Shoreline reconstructions for the Persian Gulf since the last glacial maximum. *Earth and Planetary Science Letters* **142**: 43-57.

Larminie, F. G. 1988. Silverwell Sheet (54N 02E): Sea Bed Sediments and Holocene Geology. HMSO, London.

Laraminie, F. G. 1989a. Silverwell Sheet (54N 02E): Seabed Sediments and Holocene Geology. London, HMSO.

Larminie, F. G. 1989b. Silverwell Sheet (54N 02E): Quaternary Geology. HMSO, London.

Lawley, R., and Booth, S. 2004. Skimming the Surface. *Geoscientist* **14**: 4-6.

Long, D., Wickham-Jones, C. R., and Ruckley, N. A. 1986. A flint artefact from the northern North Sea. *In* D. Roe (ed.) *Studies in the Upper Palaeolithic of Britain and North Western Europe.* BAR International Series. **296**: 55-62.

Lowe, J. J., and Walker, M. J. 1997. *Reconstructing Quaternary Environments.* 2nd Edition. Prentice Hall.

Ludwick, J. C. 1975. Tidal currents, sediment transport and sand banks in Chesapeake Bay entrance. In, L. E. Cronin, (ed.) *Estuarine Research. Volume II: Geology and Engineering*. Academic Press, New York. 365-380.

Lumsden, G. I. 1986a. *California Sheet (54N 0E): Quaternary Geology*. HMSO, London.

Lumsden, G. I. 1986b. *California Sheet (54N 0E): Seabed Sediments*. HMSO, London.

Maarleveld, T. J. and Peeters, H. 2004. Can we manage? In N. C. Flemming (ed). *Submarine Prehistoric Archaeology of the North Sea*. CBA research Report. **141**: 102-111.

Marks, B. S., and Faught, M. K. 2003. Another Prehistoric Site submerged on the continental shelf of Northwest Florida. *Current Research in the Pleistocene* **20**: (8JE1577) 49-51.

Masselink, G., and Hughes, M. G. 2003. *Introduction to Coastal Processes and Geomorphology*. Hodder Arnold.

McCave, I. N. 1971. Wave effectiveness at the sea bed and its relationship to bed-forms and deposition of mud. *Journal of Sedimentary Petrology*. **41**: 89-96.

Melis, T. S., Webb, R. H., Griffiths, P. G., and Wise, T. W. 1994. Magnitude and frequency data for historic debris flows in Grand Canyon National Park and Vicinity, Arizona. *US Geological Survey Water Resources Investigations Report*. 94-4214.

Milner, N., and Woodman, P. 2004. Looking into the canon's mouth: Mesolithic studies in the 21st century. In N. Milner and P. Woodman (eds.) *Mesolithic Studies at the beginning of the 21st Century*. Oxbow.

Milner, N. 2004. Seasonal Consumption Practices in the Mesolithic: Economic, Environmental, Social or Ritual? In N. Milner and P. Woodman (eds.) *Mesolithic Studies at the beginning of the 21st Century*. Oxbow.

Mithen, S. 1999. Hunter-gathers of the Mesolithic. In J. Hunter and I. Ralston (eds.) *The Archaeology of Britain*. Routledge. 35-58.

Mithen, S. 2003. *After the Ice*. Routledge.

Moore, P. D., Webb, J. A. and Collinson, M. E 1991. *Pollen Analysis* (Second Edition). London: Blackwell.

Morrison, A. 1980. *Early Man in Britain and Ireland*. Croom Helm.

Mueller, C., Luebke, H., Woelz, S., Jokisch, T., Wendt, G., and Rabbel, W. 2006. Marine 3-D seismic investigation of a late Ertebolle settlement in Wismar Bay (Seamap-3D case study). *Geophysical Research Abstracts* **8**.

Murray, J. W. 1987. Biogenic indicators of suspended sediment transport in marginal marine environments: quantitative examples from SW Britain. *Journal of the Geological Society, London*, **144**: 127-133.

Neubauer W., and Eder-Hinterleitner, A. 1997. 3D-Interpretation of Postprocessed Archaeological Magnetic Prospection Data. *Archaeological Prospection* **4**: 191-205.

Newell, R. C., Seiderer, L. J., and Hitchcock, D. R. 1998. The impact of dredging works in coastal waters: a review of the sensitivity to disturbance and subsequent recovery of biological resources on the seabed. *Oceanography and Marine Biology: an Annual Review*, **36**: 127-178.

Nordqvist, B. 1995. *Coastal Adaptations in the Mesolithic: A study of coastal sites with organic remains from the Boreal and Atlantic periods in Western Sweden*. GOTARC series B, **13**. Gothenburg University. Erbrink and Tacoma.

Nummedal, A. 1924. Om flintpladsene. *Norsk geologisk Tidsskrift*. Oslo. 7: 89-141.

Nygaard, S. 1990. Mesolithic Western Norway. In P. M. Vermeersch and P. Van Peer (eds.) *Contributions to the Mesolithic in Europe*. Leuven University Press. 227-237.

Oxley, I. nd. *Developments in marine Archaeological Resource management Relevant to the Revision of the Eastern Counties Research Framework*. (http://www.eaareports.demon.co.uk/FW_Oxley.pdf)

Oxley I and O'Regan D. 2001. *The Marine Archaeological Resource*. Institute Of Field Archaeologists Paper No. **4**.

(www.archaeologists.net/modules/icontent/inPages/docs/pubs/maritime_resource.pdf -)

Pedersen, L., Fischer, A., Aaby, B. (eds). 1997. *The Danish Storebaelt since the Ice Age*. A/S Storebaelt Fixed Link.

Peeters, J. H. M. 2006. *Hoge Vaart-A27 in Context: Towards a model of Mesolithic-Neolithic land use dynamics as a framework for archaeological heritage management*. Rijkedienst voor arccgeologie cultuurlanschap en monumenten. Amersfoort.

Peltier, W. R. 2004. Global Glacial Isostasy and the Surface of the Ice-Age Earth: The ICE-5G (VM2) Model and GRACE. *Ann. Rev. Earth and Planet. Sci.*, **32**: 111-149.

Paoletti, V., Secomandi, M., Priomallo, M., Giordano, F., Fedi, M., and Repolla, A. 2005. Magnetic Survey at the Submerged Archaeological Site of Baia, Naples, Southern Italy. *Archaeological Prospection* **12**: 51-59.

Posamentier, H. W. 2005. Application of 3D seismic visualisation techniques for seismic stratigraphy, seismic geomorphology and depositional systems analysis: examples from fluvial to deep-marine depositional environments. In: A. G. Dore and B. A. Vining, (eds.) *North-West Europe and Global Perspectives: Proceedings of the 6th Petroleum Geology Conference.* Geological Society of London 1565-1576.

Praeg D., 1997. Buried fluvial channels: 3D-seismic geomorphology. *In*: T. A. Davies, T. Bell, A. K. Cooper, H. Josenhans, L. Polyak, A. Solheim, M. S. Stoker, J. A. Stravers, (eds.), *Glaciated Continental Margins: An Atlas of Acoustic Images.* Chapman and Hall. 162-163.

Praeg, D. 2003. Seismic imaging of mid-Pleistocene tunnel-valleys in the North Sea Basin - high resolution from low frequencies. *Journal of Applied Geophysics* **53**: 273-298.

Price, W. A. 1963. Patterns of flow and channelling in tidal inlets. *Journal of Sedimentary Petrology.* **33**: 279-290.

Reid, C. 1913. *Submerged Forests.* Oxford.

Reynier, M. J. 1988. Early Mesolithic settlement in England and Wales: some preliminary observations. *In* N. Ashton, F. Healy and P. Pettitt (eds.) *Stone Age Archaeology: Essays in Honour of John Wymer.* Oxbow (Oxbow Monograph **102**/Lithic Studies Society Occasional Paper **6**). 174-184.

Reynier, M. J. 2005. *Early Mesolithic Britain: Origins, development and directions.* BAR British Series **393**. Archaeopress.

Rhoads, B. L., and Kentworthy, S. T. 1995. Flow structure at an asymmetrical stream confluence. *Geomorphology.* **11**: 273-293.

Roberts, A. J. 1987. Late Mesolithic occupation of the Cornish coast at Gwithian: preliminary results. *In* P. Orwley-Conwy, M. Zvelebil and H. P. Blankholm (eds.) *Mesolithic North West Europe: Recent Trends.* Department of Archaeology and Prehistory, University of Sheffield. 131-8.

Roberts, P., and Trow, S. 2002. *Taking to the Water: English Heritage's Initial Policy for The Management of Maritime Archaeology in England.* Unpublished policy statement, English Heritage, (http://www.english-heritage.org.uk/server/show/nav.8381).

Robinson, A. H. W. 1960. Ebb-flood channel systems in sandy bays and estuaries. *Geography.* **45**: 183-199.

Robinson, A. H. W. 1968. The submerged glacial landscape off the Lincolnshire coast. *Transactions of the Institute of British Geographers.* **44**: 119-132.

Rowley-Conwy, P. 1983. Sedentary hunters: the Ertebolle. *In* G. N. Bailey (ed.) *Hunter-Gather Economy in Prehistory: a European Perspective.* Cambridge University Press. 111-126.

Ryan, W., and Pitman, M. 2000. *Noah's Flood: The New Scientific Discoveries About the Event That Changed History.* Schuster.

Salomonsen, I., and Jensen, K. A. 1994. Quaternery erosional surfaces in the Danish North Sea. *Boreas* **23**: 244-253.

Shackleton, N. J. 1987. Oxygen isotopes, ice volumes and sea level. *Quaternary Science Reviews.* **6**: 183-190.

Shennan, I., Lambeck, K., Flather, R., Horton, B., McArthur, J., Innes, J., Lloyd, J., Rutherford, M., and Wingfield, R. 2000. Modelling western North Sea palaeogeographies and tidal changes during the Holocene in I. Shennan and J. Andrews (eds.) *Holocene Land-Ocean Interaction and Environmental Change around the North Sea.* Geological Society, London, Special Publication **166**: 299-319.

Shennan, I., and Horton, B. 2002. Holocene land and sea-level changes in Great Britain. *Journal of Quaternary Science* **17** (5-6): 511-526.

Sherrif, R. E. 1977. Limitations on resolution of seismic reflections and geologic detail deriveable from them. In: C. E. Payton, (ed.) *Seismic Stratigraphy - Applications to hydrocarbon Exploration.* Memoir of the American Association of Petroleum Geologists, Tulsa.

Sidell, J., Williams, K., Scaife, R., and Cameron, N. 2000. *The Holocene Evolution of the London Thames* (MoLAS Monograph 5). London: Museum of London Archaeology Service.

Smith, C. 1992. *Late Stone Age Hunters of the British Isles.* Routlege Publishers.

Smith, D. N., Roseff, R., Bevan, L., Brown, A. G., Butler, S, Hughes, G., and Monckton A. 2005. Archaeological and Environmental Investigations of a Late Glacial and Holocene river valley sequence on the River Soar, at Croft, Leicestershire. *The Holocene* **15**: 353-377.

Stewart, S. A. 1999. Seismic interpretation of circular geological structures. *Petroleum Geoscience.* **5**: 273-285.

Stride, A.H., Belderson, R.H., Kenyon, N.H. & Johnson, M.A. 1982. Offshore tidal deposits: sand sheet and sand bank facies. *In:* Stride, A.H. (ed) *Offshore tidal sands: processes and deposits.* London, Chapman and Hall, 95-125.

Stright, M. J. 1986. Evaluation of archaeological site potential on the Gulf of Mexico continental shelf using high-resolution seismic data. *Geophysics* **51** (3): 605-622.

Summerfield, M. A. 1991. *Global Geomorphology.* Longman.

Talbot, C.J. & Alavi, M. 1996. The past of a future syntaxis across the Zagros. *In:* Alsop, G.I., Blundell, D.J. & Davison, I. (eds) *Salt Tectonics.* Geological Society, London, Special Publications, **100**, 89-109.

Terrell, N., Edwards, H., Scoffield, P., and Martin, M., 2005. PGS Faroe Shetland Basin MegaSurvey – the Key to New Discoveries in a Maturing Frontier Area? In: H. Ziska, (ed.) *Faroe Islands Exploration Conference: Proceedings of the 1st Conference, Annales Societatis Scientiarum Færoensis (Faroese Society of Sciences and Humanities).* Supplementum **43**, Tórshavn. 176-182.

Tetlow, E. A. 2005. *The palaeoentomology of the salt marches and coastal woodlands of the Severn estuary.* PhD Thesis, Institute of Archaeology and Antiquity, University of Birmingham, UK.

Thomson, K. 2004. Overburden deformation associated with halokinesis in the Southern North Sea: implications for the origin of the Silver Pit Crater. *Visual Geosciences* **9**: 1-9.

Valentin, H. 1957. Glazialmorphologische untersuchungen in Ostengland: ein beitrag sum problem der letzen vereisung im Nordseeraum. *Abhandlungen des Geographischen Instituts der Freien Universität Berlin.* 4: 1-86.

Van Kolfschoten, T., and Van Essen, H. 2004. Palaeozoological heritage from the bottom of the North Sea. In N. C Flemming (ed.) *Submarine Prehistoric Archaeology of the North Sea*, CBA Research Report **141**.

Veenstra, H. J. 1965. Geology of the Dogger Bank area, North Sea. *Marine Geology.* **3**: 245-262.

Velegrakis, A. F., Dix, J. K. and Collins, M. B. 1999. Late Quaternary evolution of the upper reaches of the Solent River, Southern England, based upon marine geophysical evidence. *Journal of the Geological Society, London.* **156**: 73-87.

Verhart, L. B. M. 2004. The implications of prehistoric finds on and off the Dutch coast. *In* N. C. Flemming (ed.) *Submarine Prehistoric Archaeology of the North Sea.* CBA research Report. **141**: 57-64.

Waddington, C., Bayliss, A., Bailey, G., Boomer, I., Milner, N., Pedersen K., and Shiel R. 2003. A Mesolithic coastal site at Howick, Northumberland. *Antiquity.* **77** (295).

Ward, I., Larcombe, P., and Lillie, M. 2006. The dating of Doggerland – post glacial geochronology of the Southern North Sea. *Environmental Archaeology* **11**: 207-218.

Warren, G. 2005. *Mesolithic Lives in Scotland.* Tempus Publishers.

Warren, S. H., Clark, J. G. D., Godwin, M. E., Godwin, H., MacFadyen, W. A. 1934. An Early Mesolithic Site at Broxbourne Sealed Under Boreal Peat. *The Journal of the Royal Anthropological Institute of Great Britain and Ireland* **LXIV**: 101-128.

Watters, M. S. 2006. Geovisualization: an example from the Catholme Ceremonial Complex. *Archaeological Prospection.* **13**: 282-290.

Wenban-Smith, 2002. *Palaeolithic and Mesolithic Archaeology on the Sea Bed: Marine Aggregate Dredging and the Historic Environment.* Unpublished report for BMAPA and RCHME, Wessex Archaeology, (http://www.bmapa.org/media.htm).

Wessex Archaeology. 2003. *Artefacts from the Sea.* (http://www.wessexarch.co.uk/projects/marine/alsf/artefacts_sea/artefacts_sea.html).

Wessex Archaeology. 2005. *England's Historic Seascapes. Historic Environment Characterisation in England's Intertidal and Marine Zone.* (http://www.english-heritage.org.uk/server/show/nav.8684).

Wickham-Jones, C. R., and Dawson, S. 2006. *The scope of Strategic Environmental Assessment of North Sea areas SEA7 with regard to prehistoric and early historic archaeological remains.* Department of Trade and Industry. (http://www.offshore-sea.org.uk/site/scripts/category_info.php?categoryID=37)

Whitehouse, N. J., and Smith, D. N. 2004. "Islands" in Holocene forests: implications for forest openness, landscape clearance and "culture-steppe" species. *Environmental Archaeology.* **9**(2): 203-212.

Wingfield, R. T. R. 1990. The origin of major incisions within the Pleistocene deposits of the North Sea. *Marine Geology.* **91**: 31-52.

Wright, L. D., Coleman, J. M., and Thom, B. G. 1975. Sediment transport and deposition in a macrotidal river channel: Ord River, Western Australia. In, L. E. Cronin, (ed.) *Estuarine Research. Volume II: Geology and Engineering.* Academic Press, New York. 309-321.

Wymer, J. 1991. *Mesolithic Britain.* Princes Risborough: Shire Publications.

Mapping Doggerland

The prehistoric landscapes of the North Sea basin are amongst the most enigmatic archaeological landscapes in northwestern Europe. Whilst the region contains one of the most extensive and, probably, best preserved hunter-gatherer landscapes in Europe, global warming resulted in the loss of a vast area of habitable land over a period of c.11,000 years. The challenge to investigate, interpret and manage the heritage of this extraordinary, but largely inaccessible landscape is enormous. 3D seismic datasets, acquired to explore deep geology, present a major opportunity to explore this landscape at a regional scale. The North Sea Palaeolandscapes Project utilised c. 23,000 km^2 of 3D seismic data to provide detailed digital mapping of Late Pleistocene and Holocene topographic features across the area. This publication provides an assessment of the archaeological potential of the interpreted landscape and related environmental sources. It outlines associated archaeological issues, a methodology for implementing historic landscape characterisation within the area, as well as an assessment of data sources for further exploration of the British coastal shelf. The results of this study will be of interest to archaeologists, geomorphologists and cultural resource managers working in analogous environments, whilst the methodology outlined may be applied to similar landscapes with comparable supporting data.

La Cartographie de Doggerland

Les paysages préhistoriques du bassin de la Mer du Nord sont parmi les plus énigmatiques de l'Europe du Nord-Ouest. Tout en admettant que la région possède un des plus vastes paysages, et selon toute probabilité, le paysage le mieux conservé des chasseurs-cuilleurs en Europe, le réchauffement de la planète a entraîné la perte d'une région énorme de terrain habitable à travers une période d'environ 11,000 ans. Le défi énorme est d'étudier, d'interpréter, et de gérer l'héritage de ce paysage qui est à la fois extraordinaire et, pour la plupart, inaccessible. Des données sismiques à trois dimensions qui ont été acquises pour explorer la géologie profonde, nous offrent une opportunité importante d'explorer ce paysage sur un plan régional. Le Projet des Paléo-paysages de la Mer du Nord a utilisé environ 23,000 km2 de données sismiques en relief afin de créer une carte digitale détaillée des caracteristiques topographiques de la fin de l'ère Pleistocene et Holocene de la région. Cette publication fournit une évaluation des possibilités archéologiques du paysage interprété et des sources environnementales. Elle met en relief des questions archéologiques, une méthodologie pour réaliser la caractéristaion des paysages historiques dans ce domaine, et une évaluation des sources des données pour explorer d'avantage l'écueil anglais. Les résultats de cette étude intéresseront des archéologistes, des géomorphologistes, et des directeurs de ressources culturelles qui travaillent dans des environnements pareils, et, on pourra utiliser la méthodologie soulignée avec d'autres paysages qui sont comparables et qui ont des données similaires.

Die kartographische Aufnahme von 'Doggerland'

Die prähistorischen Landschaften unter der Nordsee gehören zu den rätselhaftesten archäologischen Landstrichen des nordwestlichen Europas. Obgleich die Region eine der weiträumigsten und, wahrscheinlich, eine der best erhaltensten Landschaften prähistorischer Jäger und Sammler in Europa umfaßt, hat die Erwärmung der Erdatmosphäre über einen Zeitraum von ungefähr elftausend Jahren zum Verlust einer ausgedehnten Fläche bewohnbaren Landes geführt. Die Herausforderung, diese außergewöhnliche aber größtenteils unerreichbare Landschaft zu untersuchen, zu interpretieren und ihre Hinterlassenschaft zu bewahren, ist enorm. Seismische 3D Datenbestände, die zur geologischen Erforschung tieferer Schichten angelegt wurden, bieten uns eine einzigartige Gelegenheit, diese Landschaft auf regionaler Ebene zu erkunden. Das Nordsee Paläo-Landschafts Projekt hat seismische 3D Daten auf einer Fläche von ungefähr 23 000 km^2 genutzt um eine detaillierte Karte der topographischen Merkmale dieses Gebietes während des späten Pleistozäns und des Holozäns herzustellen. Dieses Buch bietet eine Beurteilung des archäologischen Potentials dieser Landschaft und eine Einschätzung der verwandten Quellen zur Umwelt. Es umreißt die damit verbundenen archäologischen Fragen, skizziert eine Methodik für die Erfassung des Charakters historischer Landschaften in diesem Gebiet und beurteilt andere Arten von Materialien für die weitere Erforschung des Meeresbodens vor der britischen Küste. Die Ergebnisse dieser Studie sind für alle Archäologen, Geomorphologen und Leiter kultureller Ressourcen, die auf diesem und ähnlichem Gebiet tätig sind, von Interesse, während die hier vorgeschlagene Methodik auch auf andere Landschaften dieser Art, für die vergleichbare Daten zur Verfügung stehen, angewandt werden kann.

Index

0-9

2D seismics lines, 7
3D seismic survey, 11, 13, 93, 94, 96, 97, 103, 108

A

Acetylation, 99
Acoustic energy, 23
Acoustic impendence contrasts, 23, 24, 27
Archaeological survey, 35
Archaeology Data Service, 67
Attribute analysis, 31, 34
Avizo, 38, 39
Azimuth, 28

B

Bandwidth, 25, 30, 33
Barrier Islands, 52, 53
Bathymetry, 4, 23, 31, 49, 51, 56, 82, 108, 114
Binned dataset, 28
Borehole control, 28
Borehole logs, 12
Botney Cut, 49, 51, 82, 86, 88, 112
Bristol Channel, 16, 17
British Geological Survey, 11, 15, 67, 93
Broxbourne, 105

C

CAD, 40
Climate change, 9, 11, 105
Coastal zone, 5, 16
Coherence, 28, 29, 30
Coherence cube, 28
Coles, Bryony, 1, 3, 4, 8, 75, 82, 93
Colluvium, 51
Communications, 1
Continental shelf, 1, 3, 6, 8, 12

D

Data audit, 8, 16, 22
Delta system, 82
Depositional facies, 28
Depositional systems, 7, 28
Digital mapping, 6, 67
Dogger Bank, 4, 29, 51, 74, 75, 112
Doggerland, 1, 2, 3, 4, 105, 118

E

Early Holocene, 23, 75, 82, 90, 93, 116
Elbow Formation, 51, 82, 97
English Channel, 13, 30
English Heritage, 6, 8, 13, 14, 110, 111
Environmental potential, 8, 93, 97, 108

Eustasy, 51

F

Fakespace PowerWall, 34
Flemming, Nic, 1, 3, 4, 8, 75, 93, 108, 110
Floodplains, 82, 89
Fluvial channels, 28, 81, 82, 112
Fluvial processes, 51
Fluvial systems, 4, 51, 75, 78, 86, 112
Fluvial valleys, 4
Foresets, 49
Fresnel Zone, 25, 30

G

Geographic Information Systems, 33, 35
Geological permeability, 74
Geology, 15, 75, 81
Geophysical survey, 6, 8, 11, 12, 30, 41
Geophysics, 12, 15
GIS, 11, 13, 14, 15, 16, 34, 35, 37, 40, 41, 67
Glacial, 1, 8, 13, 50, 78, 82, 89, 91, 112
Glacio-lacustrine deposits, 82
Global warming, 8
Google Earth, 67
Graben collapse, 72, 73, 107
Gravimeter, 12
Great North Sea plain, 117
Greater Wash, 21
Ground penetrating radar, 38
Ground truthing, 8, 16, 59, 90

H

Heritage, 1, 6, 8, 13, 106, 108, 109, 110, 116, 118
High resolution 2D seismic, 4
High resolution 3D seismic, 4
Hinterland, 11
Holocene, 1, 3, 4, 5, 6, 8, 9, 11, 22, 23, 24, 28, 29, 30, 33, 43, 49, 50, 51, 52, 53, 54, 55, 56, 58, 67, 69, 70, 71, 72, 73, 74, 75, 77, 78, 79, 81, 82, 83, 86, 87, 89, 90, 93, 97, 101, 103, 105, 106, 108, 109, 110, 112, 114, 116, 117
Horsham, 105
Howick, 105
HP VISTA, 33, 34
Hunter gatherer, 1, 78, 106
Hydraulic energy, 52
Hydrocarbon exploration, 23, 24, 25, 28, 30, 31, 35
Hydrocarbon resources, 11
Hydrophone, 12

I

Image quality, 34
Inner Silver Pit, 86
Institute of Archaeology and Antiquity, 33
Internet mapping, 67

Inundation, 51, 56, 72, 74, 75, 93, 105, 108
Isostasy, 51
Isostatic models, 75
Isostatic rebound models, 4
Isosurfacing, 38

K

KOH digestion, 99

L

Late Holocene, 4, 56, 93
Late Pleistocene, 5, 6, 8, 33, 51, 74, 75, 82, 86, 93, 101, 103, 112
Late Quaternary, 1, 6
Lateral resolution, 25, 28, 30
Lithological predictions, 7
Littoral sediment transport convergence, 56

M

Macrofossil, 97, 98
Magnetometer, 12
Magnetometry, 35
Mapped horizons, 28
Marine aggregates, 13
Marine archaeology, 35, 110
Marine environment, 16, 22, 72, 110
MESH, 15
Mesolithic, 1, 6, 8, 16, 21, 28, 30, 72, 105, 106, 111, 112, 114, 117
Mesozoic Coast, 21
Middle Pleistocene, 81, 103
Moorlog, 1, 21
Morphological features, 30
Morphology, 4, 27, 44, 51, 52, 53, 56, 58

N

National Geosciences Data Centre, 12
National Hydrocarbons Data Archive, 12
Neolithic, 17
Normalisation, 116
Norwegian Trough, 108

O

Opacity filters, 28
Opacity rendering techniques, 28, 30, 34
Outer Silver Pit, 4, 7, 43, 44, 52, 55, 75, 81, 103, 104, 112, 116
Overburden, 24

P

Palaeochannels, 81, 85, 86, 93
Palaeocoastline, 7
Palaeoenvironmental data, 30, 106
Palaeogeography, 4, 6
Palaeohydraulic reconstruction, 59
Palaeolandscape, 4, 16, 81
Palaeolithic, 16, 89, 108, 109, 110, 118

Periglacial, 50
PGS MegaSurvey, 21
Picking techniques, 38
Planar slices, 40
Pleistocene, 6, 24, 50, 72, 74, 75, 81, 82, 86, 103, 112
Pollen, 97, 99, 101
Portland, 21
Prehistoric coastlines, 16
Prehistoric landscapes, 1, 8, 116

Q

Quaternary, 5, 43, 49, 51, 67

R

Race Bank, 21
Radiometric dating, 101, 102
Reflection coefficient, 23
Reid, Sir Clement, 1
Remotely sensed imagery, 35
Research strategies, 106
Resistivity, 35
Resolution, 4, 8, 11, 12, 13, 21, 23, 24, 25, 28, 30, 34, 49, 55, 59, 67, 75, 82, 86, 90, 101, 106, 108, 111
Rheology, 51
Root mean squared, 35

S

Salt dome, 72, 73, 86
Salt marsh, 82, 85, 112
Sand banks, 55
Sand waves, 53, 55, 58, 75, 82, 89
Scaleable visualisation system, 34
Seabed reflector, 29, 30
Seabed samples, 23, 31
Seabed sampling, 4
Seabed topography, 4, 23, 37
Sea-level rise, 78
Sediment core analysis, 51
Sediment sampling, 11, 13, 15, 16, 102
Sediment structure, 23
Sedimentary layers, 13
Sedimentation, 15
Sedimentological analysis, 58
SEG-Y data, 38
Seismic amplitude, 28, 29
Seismic attribute analysis, 7, 28
Seismic coherence, 28
Seismic profile, 23, 27
Seismic reflection, 12, 23, 24, 25, 26, 27, 30
Seismic reflection profiles, 23
Seismic slices, 28
Seismic surveys, 6, 12, 14, 21, 35, 108
Seismic visualisation, 31
Seismic volume, 27, 28, 29, 37
Semblance, 28, 29
Shallow core, 11, 23, 31, 97
Shallow coring, 4, 23
Shotton River, 78, 102, 103
Silverwell, 50

SMT Kingdom, 35
Solid modelling, 34, 36, 38, 40, 41
Sonar, 11, 12, 16, 21, 59
Spatial analysis, 40, 41
Spits, 52, 53
Spurn, 21, 112
Star Carr, 105
Stereo visualisation, 36
Stereoscopic display, 34
Storegga landslide, 93
Stratigraphy, 11, 12
Sub-bottom seismic profiling, 12, 21
Subglacial processes, 43
Swarte Bank, 81, 112

T

Talus, 51
Tershellingerbank, 50
Thatcham, 105
Threat mapping, 116
Tidal asymmetry, 57, 58
Tidal scour marks, 78, 82, 112
Tigress, 35, 40
Timeslices, 28, 35, 67, 75
Topographic data, 6, 56, 111
Trace spacing, 25, 27, 30

Triton Knoll, 21
Tuning thickness, 24
Tunnel valley, 51, 78, 82, 86

U

UK Onshore Geophysics Library, 12
Uncertainty Mapping, 116
University of Birmingham, 6, 33

V

Vibrocore, 97, 101, 103, 104, 108
Viking Bank, 108, 109
Viking Bergen Hills, 108
Voxel rendering, 6
Voxel volume, 28, 40, 41

W

Weichselian, 4, 49, 82
Weichselian maximum, 4
Well Hole, 86, 88, 112
WesternGeco, 21
Wetlands, 78, 93, 102, 106, 107, 112
Wytch Farm, 30